No.1 2013

世界園林
WORLDSCAPE

中国林业出版社

本辑主题：填埋场复育景观

THEME: THE LANDSCAPE OF LANDFILLS

ARIEL SHARON PARK AND HIRIYA LANDFILL TRANSFORMATION, ISRAEL

以色列阿瑞尔夏隆公园及海瑞亚填埋场的质变

MASTER DIALOGUE: PETER LATZ

对话大师：彼得·拉茨

INTRODUCTION OF PARTIAL NEW CULTIVARS OF BEDDING FLOWERS

草花育成品种介绍

ndscape Competition carry out annually.
g and Macao regions meet the conditions of

Awards: Residential landscape award is divided into Comprehensive, design, project and innovation awards.

Time schedule: The works declare time is from June to October

Sponsor: International Association of Green Architecture and Residential Landscape
Chinese Flowers Gardening and Landscaping Industry Association

Organizer: www.chla.com.cn
Worldscape Magazine
COHT China Forestry Publishing House

Contact Us:

Website: http://db.chla.com.cn/residence/
Tel: (86) 010-88360800 88366270 88361443
Fax: (86) 010-88365357
E-mail: chla2012@126.com
Address: 1409#, Ganjiakou building, No.17, Sanli River Road, Haidian District, Beijing.
Post Code: 100037

图片提供：北京源树景观规划设计事务所/北京龙湖置业有限公司

"园冶杯"风景园林国际竞赛

2012

"园冶杯"风景园林（毕业作品、

The 2012 "Yuan Ye Award" International Landscape Architecture Grad

"园冶杯"风景园林（毕业作品、论文）国际竞赛是由中国建设教育协会和中国花卉园艺与园林绿化行业协会主办，中国风景园林网和《世界园林》杂志社承办，在风景园林院校毕业生中开展的一项评选活动。

The 2012 "Yuan Ye Award" International Landscape Architecture Graduation Project is hosted by China Construction Education Association, Chinese Flowers Gardening and Landscaping Industry Association, and is undertaken by China landscape architecture network and the magazine of Worldscape. It is held in landscape architecture school for graduates to contest.

时间安排：

报名截止日期2012年4月30日；资料提交截止日期2012年6月15日

参赛资格：

应届毕业生（本科、硕士、博士）

参赛范围：

风景园林及相关专业的毕业作品、论文均可报名参赛

竞赛分组：

竞赛设置四类：风景园林设计作品类、风景园林规划作品类、
园林规划设计论文类、园林植物研究论文类。

每类设置两组：本科组和硕博组。

地址:北京市海淀区三里河路17号甘家口大厦1409　邮编:100037　电话:（86）010-88364851/（86）010-88361443　传真:（86）010-88365357 / （86）010-88361443

学生咨询邮箱:yyb@chla.com.cn　院校咨询邮箱:messagefj@126.com　官方网站:http://chla.com.cn/

论文）国际竞赛

uation Project/Thesis Competition

Time schedule:

The application deadline is April 30, 2012; Submission deadline on May 31, 2012

Qualification:

This year's graduates (Bachelor, Master, Doctor)

Competition content:

Landscape architecture and the related specialized graduation work, or paper

Types:

There are four types: The design works of landscape architecture, the planning works of landscape architecture, the design and planning papers of landscape architecture, and the papers of garden plants. Every type is set into two groups: the undergraduate group, the master and doctor group.

Contact Us:

Address: 1409#, Ganjiakou building, No.17, Sanli River Road, Haidian District, Beijing.

Post Code: 100037

Tel: 010-88360800 88366270 88361443

Fax: 010-88365357

Student Advisory E-mail: yyb@chla.com.cn

College Advisory E-mail: messagefj@126.com

Website: http://chla.com.cn/

世界园林

WORLDSCAPE

主办单位 中国花卉园艺与园林绿化行业协会
国际绿色建筑与住宅景观协会
出版单位 中国国际园林集团公司

（按姓氏字母顺序排名）
总　编 王小璘（台湾）
副总编 包满珠　李　敏　刘滨谊　沈守云　王　浩　周　进　朱育帆
顾问编委 凌德麟（台湾）　罗哲文
编委会
常务编委 Jack Ahern（美国）　曹南燕　陈蓁蓁　高　翅　Christophe Girot（瑞士）
Karen Hanna（美国）　何友锋（台湾）　贾建中　况　平　Eckart Lange（英国）
李如生　李　雄　李炜民　刘滨谊　Patrick Miller（美国）欧圣荣（台湾）
强　健　Phillippe Schmidt（德国）Alan Tate（加拿大）　王庚飞　王良桂　王向荣
谢顺佳（香港）杨重信（台湾）　喻肇青（台湾）章俊华　张　浪　赵泰东（韩国）
周　进　朱建宁　朱育帆
编　委 白祖华　陈其兵　成玉宁　杜春兰　方智芳（台湾）　黄　哲　简仔贞（台湾）
金晓玲　李春风（马来西亚）李建伟　李满良　林开泰（台湾）刘纯青　刘庭风
罗清吉（台湾）马晓燕　Hans Polman（荷兰）　邱坚珍　瞿　志　宋钰红
王明荣　王鹏伟　王秀娟（台湾）　吴静宜（台湾）　吴雪飞　吴怡彦（台湾）
夏海山　夏　言　张莉欣（台湾）　张青萍　周武忠　周应钦　朱　玲　朱卫荣
郑占峰
编辑中心 北京交通大学建筑与艺术学院
主　任 夏海山
副主任 金晓玲　刘纯青　蒙小英　吴雪飞　朱　玲
责任编辑 蒙小英
编　辑 陈　鹭　傅　凡　高　杰　孟　彤　佘高红　张红卫　赵彩君
外文编辑部 何友锋（台湾）Charles Sands（加拿大）（主任）Trudy Maria Tertilt（德国）
谢顺佳（香港）　朱　玲
美　编 张　野
排版设计 王　薇
编辑部 梁兴芳　郑维伟（主任）
国际部 马一鸣　高　杰　宋焕芝　覃　慧（台湾）　郑晓笛（主任）

地　址
北　京 北京市海淀区三里河路 17 号甘家口大厦 1409
邮 编：100037　电 话：86-10-88364851
传 真：86-10-88361443　邮 箱：worldscape@chla.com.cn
香　港 香港湾仔骆克道 315-321 号骆中心 23 楼 C 室
电 话：00852-65557188　传 真：00852-31779906
台　湾 台北书局
台北市万华区长沙街二段 11 号 4 楼之 6
邮 编：108　电 话：886+2-23121566,
传 真：886+2-23120820　邮 箱：nkai103@yahoo.com.tw
编辑中心 北京市海淀区西直门外北京交通大学建筑与艺术学院 207 室
邮 编：100044　电 话：86-10-51684910
传 真：86-10-51684910　邮 箱：Worldscape_c@chla.com.cn

封面作品 阿瑞尔夏隆公园眺望亭（图片来源：Latz+Partner）

图书在版编目（CIP）数据

世界园林 . 填埋场复育景观 : 汉英对照 / 中国花卉园艺与园林绿化行业协会主编 .
-- 北京 : 中国林业出版社 , 2013.3
ISBN 978-7-5038-7039-2
Ⅰ . ①世… Ⅱ . ①中… Ⅲ . ①卫生填埋场 – 园林设计 – 景观设计 – 作品集 – 世界 – 汉、英 Ⅳ . ① TU986.61
中国版本图书馆 CIP 数据核字 (2013) 第 091909 号

中国林业出版社
责任编辑：李　顺　成海沛
出版咨询：（010）83223051

出　版：中国林业出版社（100009 北京西城区德内大街刘海胡同 7 号）
印　刷：北京卡乐富印刷有限公司
发　行：中国林业出版社
电　话：（010）83224477
版　次：2013 年 3 月第 1 版
印　次：2013 年 3 月第 1 次
开　本：889mm x 1194mm 1 / 16
印　张：12.5
字　数：200 千字
定　价：定 价：80.00RMB（30USD，150HKD）

Host Organizations
China Hortiflora and Landscaping Industry Association
International Association of Green Architecture and Residential Landscape
Publisher China International Landscape Group Limited

Editor-in-Chief
Xiaolin Wang（Taiwan）
Deputy Editors
Manzhu Bao　Min Li　Binyi Liu　Shouyun Shen　Hao Wang　Jin Zhou　Yufan Zhu
Consultants
Delin Ling（Taiwan）　Zhewen Luo
Editorial Board
Managing Editors
Jack Ahern（USA）　Nanyan Cao　Zhenzhen Chen　Chi Gao　Christophe Girot（Switzerland）
Karen Hanna（USA）　Youfeng He（Taiwan）　Jianzhong Jia　Ping Kuang　Eckart Lange（England）
Rusheng Li　Xiong Li　Weimin Li　Binyi Liu　Patrick Miller（USA）　Shengrong Ou（Taiwan）　Jian Qiang
Phillippe Schmidt（Germany）　Alan Tate（Canada）　Geng fei Wang　Lianggui Wang　Xiangrong Wang
Shunjia Xie（HongKong）　Chongxin Yang（Taiwan）　Zhaoqing Yu（Taiwan）　Junhua Zhang
Lang Zhang　Taidong Zhao（Korea）　Jin Zhou　Jianning Zhu　Yufan Zhu
Senior Editors
Zuhua Bai　Qibing Chen　Yu-ning Cheng　Chunlan Du　Zhifang Fang（Taiwan）
Zhe Huang　Yuzhen Jian（Taiwan）　Xiaoling Jin　Chunfeng Li（Malaysia）　Jianwei Li
Manliang Li　Kaitai Lin（Taiwan）　Chunqing Liu　Tingfeng Liu　Qingji Luo（Taiwan）
Xiaoyan Ma　Hans Polman（Netherlands）　Jianzhen Qiu　Zhi Qu　Yuhong Song
Mingrong Wang　Pengwei Wang　Xiujuan Wang（Taiwan）　Jingyi Wu（Taiwan）
Xuefei Wu　Yiyan Wu（Taiwan）　Haishan Xia　Yan Xia　Jianwei Ling
Lixin Zhang（Taiwan）　Qingping Zhang　Wuzhong Zhou　Yingqin Zhou　Ling Zhu
Weirong Zhu　Zhanfeng Zheng
Editorial Office
School of Architecture and Design, Beijing Jiaotong University
Editorial Director
Haishan Xia
Deputy Editorial Director
Xiaoling Jin　Chunqing Liu　Xiaoying Meng　Xuefei Wu　Ling Zhu
Executive Editor
Xiaoying Meng
Editorial Assistants
Lu Chen　Fan Fu　Jie Gao　Tong Meng　Gaohong She　Hongwei Zhang　Caijun Zhao
Foreign Language Editorial Department
Youfeng He (Taiwan)　Charles Sands（Canada, Director）　TrudyMaria　Tertilt（Germany）
Shunjia Xie（Hongkong）　Ling Zhu
Art Editor
Ye Zhang
Layout Design
Wei Wang
Editorial Department
Xingfang Liang　Weiwei Zheng（Director）
International Department
Yiming Ma　Jie Gao　Huanzhi Song　Hui Qin (Taiwan)　Xiaodi Zheng（Director）

Corresponding Address
Beijing　1409A Room, Gan Jia Kou Tower, NO. 17 San Li He Street, Haidian District, Beijing P.R.C
Code No. 100037　Tel: 86-10-88364851　Fax: 86-10-88361443　Email: worldscape@chla.com.cn
HongKong Flat C,23/F,Lucky Plaza,315-321 Lockhart Road, Wanchai, HONGKONG
Tel: 00852-65557188　Fax: 00852-31779906
Taiwan　Taipei Bookstore
6#,the 4th Floor , Changsha Street Section No.2, Wanhua District , Taipei Code No. 108
Tel: 886+2-23121566　Fax: 886+2-23120820　Email: nkai103@yahoo.com.tw
Editing Center　207#,School of Architecture and Design,Beijing Jiaotong University　Code No. 100044
Tel: 86-10-51684910　Fax: 86-10-51684910　Email: Worldscape_c@chla.com.cn

Publishing Date March 2013

Cover Story ARIEL SHARON PARK（Source:Latz+Partner）

WORLDSCAPE 目录

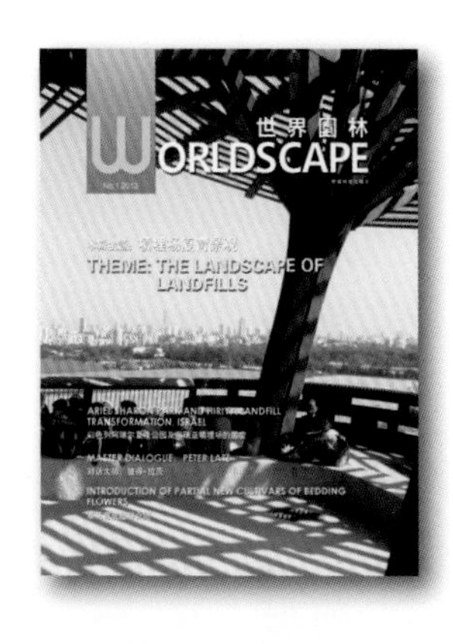

世界园林 第二辑
主　　题 填埋场复育景观

WORLDSCAPE
No.1 2013
THEME: THE LANDSCAPE OF LANDFILLS

北京房山万亩滨水森林公园

北京北小河公园

北京植物园展览温室

北京长安街环境营造

北京紫竹院公园青莲岛

CONTENTS

WORLDSCAPE

新近设计落成：万达长白山国际度假区

总编心语 EDITORIAL

王小璘
Xiaolin Wang

自从20世纪以来，垃圾场、采石场及矿场的填埋后修复再利用一直是许多专业领域所关注的议题。由于它涉及的问题十分广泛，因此无论废弃物的分解、填埋气体的回收和再利用、庞大的土方工程和植物群落的生长和稳定等，都考验着人类的智慧。对于景观园林师而言，垃圾填埋场的复育和再利用不仅是科技和自然的结合，也是文化和自明性的表征。

诸多成功的实例显示，经过完善的调查研究以及规划设计和施工管理，一个营运数十年的垃圾填埋场可以化腐朽为神奇地成为一个城市休闲公园、风景游憩地、自然保护区、农业生产地，甚至森林、湿地及野生动植物栖息地，进而成为生态平衡、科研和环教的重要场所。

由彼得·拉茨联合事务所设计的以色列阿瑞夏隆公园和海瑞亚填埋复育，彰显了水利工程和景观园林高度的专业知识和技术，它成功地将一个垃圾填埋场转化为公园，并将一个受洪水侵犯的平原妥善规划而保留了它的机能。Niall Kirkwood教授的专文体现了这个公园如何带给人们一个不同凡响的体验和感受。尼尔班森介绍了英国达勒姆遗址海岸如何以崭新的手法达成满足21世纪需求的海岸质量，透过其描述长期参与规划和管理工作，使读者有着身临其境的奇妙氛围。上海辰山植物园矿坑花园面积仅约4.3hm²，然而，也曾遭受生态环境的严重破坏。面临诸多挑战，朱育帆教授及其设计团队同时用"加减法"进行采石矿坑特殊环境的生态修复，进而建设成为一个精致的、有特色的修复式花园。丹麦斯缇岛原是一个使用过27年的垃圾填埋场，经由完善的修复工程，使得场地成为一个提供附近城市居民多元功能的休憩场所。高雄都会公园位于台湾南部，其二期工程由汪荷清总监及其设计团队所规划，并由该公园管理站洪钦勋主任负责全园经营管理及维护工作。公园原是一个占地95hm²的垃圾填埋场，于使用封闭后经过多年的复育工作，如今已是高雄都会区市民最佳的休闲游憩场所和各种生物的栖息地。清溪垃圾填埋场公园是纽约市最大的公园，由世界最大的垃圾填埋场改造而成，它不仅提供大都会地区健康的生态系统和可持续能源再利用的机制，更创造出可供多元使用的场所，为拥挤的都会区营造一个可呼吸的空间。郑晓笛博士以亲临现场的经验道出了此一耗时30年兴建的景观工程，如何成为众所瞩目的焦点。

本期焦点人物彼得·拉茨教授拥有多元和丰富的景观专长和经验，在填埋场复育景观方面无疑是国际最知名的景观大师。由本刊引述十位知名的景观专家学者的话语中，读者可以洞察拉茨先生如何引领当代景观设计思潮。而本刊也特别邀请到拉茨教授在「大师论坛学术报告会」上与多位景观专家学者进行历史性的对谈，相信这些信息都将带给读者深刻的启示！

众所周知，园林植物选种是景观规划设计中最重要的一环，由于它丰富多样的种类而充实了人类的生活环境。本期介绍了5种由包满珠教授率领的研究团队培育出的新品种。

最后，本期介绍4个竞图获奖作品，以飨读者。□

王小璘

Since the latter part of the last century the transformation and reuse of garbage dumps, quarries and pit mines has been a key issue for various professions including landscape architecture. And because of the wide range of technologies required for related practices such as waste decomposition, landfill gas collection, methane harvesting, landfill and earthwork construction, green infrastructure implementation, and strategized planting, human wisdom and understanding has been challenged. For a landscape architect, the restoration of a landfill site does not only combine technology and nature, it also represents the local culture and identity.

Many successful projects have shown that through dedicated survey, investigation, research, planning, design, construction and management, a landfill which operated over a period of centuries can be transformed into an urban leisure park, a scenic recreation area, a natural protection area, productive agricultural land, or even a forest, wetland or wildlife habitat. By doing so, multiple functions of ecological balance benefit the environment and aid in scientific study and environmental education.

Ariel Sharon Park and Hiriya Landfill Park by Latz & Partner are projects that demonstrate highly specialized technical knowledge of drainage and hydraulics as well as innovative landscape design. They successfully transform a landfill site into a park and preserve the function of a flooded plain through an impressive planning process. The article by Professor Niall Kirkwood reveals how this park brings visitors a remarkable feeling and experience. Mr. Niall Benson introduces how a severely polluted coastal landscape in England was brought back to life and has been repositioned for 21st century needs. Through the description of the author, who is involved in long-term planning and management work, the reader will experience an inspirational atmosphere. Shanghai Chenshan Botanic Garden covers 4.3ha of an area that was previously in a state of severe ecological decay. Professor Yufan Zhu and his design team adopted the strategy of "Addition and Subtraction" as the design principle for the ecological reclamation of this special form of quarry. They have transformed the site into a delicate and distinctive garden. Stige Island, was used as a landfill site for 27 years. Through the transformation plan, it has become a recreational area, providing multiple functions for the visitors from the vicinal cities in Demark. Kaohsiung Metropolitan Park (KMP), located in the south of Taiwan, has completed the second stage of a reclamation project for three years. Through the design of Ms. Herching Wang, Director of the Cosmos Inc. Planning & Design Consutants., and the management of Mr. Chinshun Hon, Director of Management Section of KMP, a 95ha area originally used as a landfill, has been transformed into not only a diverse biological habitat for wildlife, but also one of the best recreational areas for the citizens of Kaohsiung city. Professor James Corner transformed the world's largest landfill into the largest public park in the New York metropolitan area. It not only provides a healthy eco-system and a sustainable dynamic energy reuse system, but also a multi-functional breathing space in a crowded city. Dr. Zheng's article addresses the unique planning process of this project.

Although the work of Professor Peter Latz extends beyond landfill reclamation, he is undoubtedly the most internationally well-known landscape architect working in this field. The reader can recognize from the comments of ten internationally outstanding scholars and landscape architects how Peter has led the way in new trends of contemporary landscape design. The reader can also gain insight into some of Peter's personal opinions and learn about his poignant design philosophy in the 'Master Dialogue' section.

As we all know, plant selection is the most important aspect of landscape planning and design. An abundant diversity of plant species enriches our human living environment. In this issue we recommend five new varieties cultivated by a research group led by Professor Manzhu Bao.

And finally, we introduce four competition winning works for our readers. ■

2012年度法国年轻景观师和建筑师奖

近日，法国文化部公布了17位获得2011–2012年度“法国杰出年轻景观师和建筑师奖(AJAP)”的名单。该奖项设立于1980年，每两年一届，由法国文化和宣传部组织评选年龄小于35岁的优秀建筑设计师，并从2006年起加入景观设计师奖项，是法国年轻景观建筑类设计师的最高荣誉。本次获奖者包括了14个建筑师团队和3个景观师团队，由法国前文化部长傅雷德里克·密特朗为首的评委会选出。众多评委中不乏业界的国际大师，如当代著名的景观设计大师、2011年的欧洲规划大奖和2012年的法国景观大奖的获得者——米歇尔·德文涅。

米歇尔·德文涅，1958年生，1984年毕业于法国凡尔赛国立高等景观学院，并取得法国国家景观建筑师文凭(Paysagiste DPLG)。1988年在巴黎成立其事务所，其工作范围除了项目设计，还包括为一些公立和某些私人机构的课题调查和研究，其参与的项目遍布全球，多次领衔设计与国际建筑大师（理查德·乔治·罗杰斯、伦佐·皮亚诺、赫尔佐格和德梅隆、雷姆·库哈斯、诺曼·福斯特、让努·维尔等）合作的欧洲和美国的多项城市改造和设计项目，他的出现真正有力地提升了景观设计师在城市规划领域的地位，使景观设计师排在城市规划师和建筑师之前。

“过去的30年，景观师的任务只是给城市做一些装饰，但今天他们参与到了城市以及领土的更新和重组。”米歇尔·德文涅如是说。这些年轻的设计师用他们的创造力给行业带来了一股新鲜的空气。

第160届里昂灯光节——城市夜景的饕餮盛宴

里昂灯光节始于1852年，于每年的12月8号开始并持续4天。灯光节最初完全是由市民自发组织的节日，每家每户都会在窗台上摆满蜡烛，为了纪念和感谢圣母玛丽亚对里昂城的保护。1989年后，这项自发的节日开始转变为由里昂政府出资组织，众多布

景艺术家参与的公众节日，丰富多样的灯光布景和投影被应用于节日期间，里昂灯光节也因此每年吸引着大量的游客前来观看，并逐渐成为一张城市的特色名片。近年来由于公共艺术对城市空间的大量介入，灯光节为景观设计师和城市照明设计师以及当代艺术家提供了一个合作的和发挥创造力的契机。2012年12月的第160届灯光节，许多临时性的景观灯光作品结合里昂古色古香的老

建筑在节日期间被展示出来，一个完全不同的城市在夜色中呈现在人们面前，整个里昂城就像一个超大的布景的活力四射的布景舞台。（部分摄影：Fabrice Dimier，Muriel Chaulet，Michel Djaoui）

新加坡欲打造世界植物之都“超级树花园”可见一斑

2012年圣诞节之际，新加坡滨海湾公园内12棵人造太阳能“超级树”在夜幕降临后缤纷闪烁，营造出浓厚的节日气氛，成为公园内的核心景观。新加坡政府计划在滨海湾公园内总共建造3个花园，欲打造世界植物之都，超级树花园仅是其中之一。据悉，“超级树花园”的建造共耗资3.5亿英镑（约合人民币35亿元）。

这些超级树位于新加坡滨海湾公园的核心区域，高达25m至50m。树上装配了太阳能电池板和雨水收集器，既能为霓虹灯提供电力，又可为藏身树干中的空中花园提供灌溉用水。空中花园中种植着来自世界各地

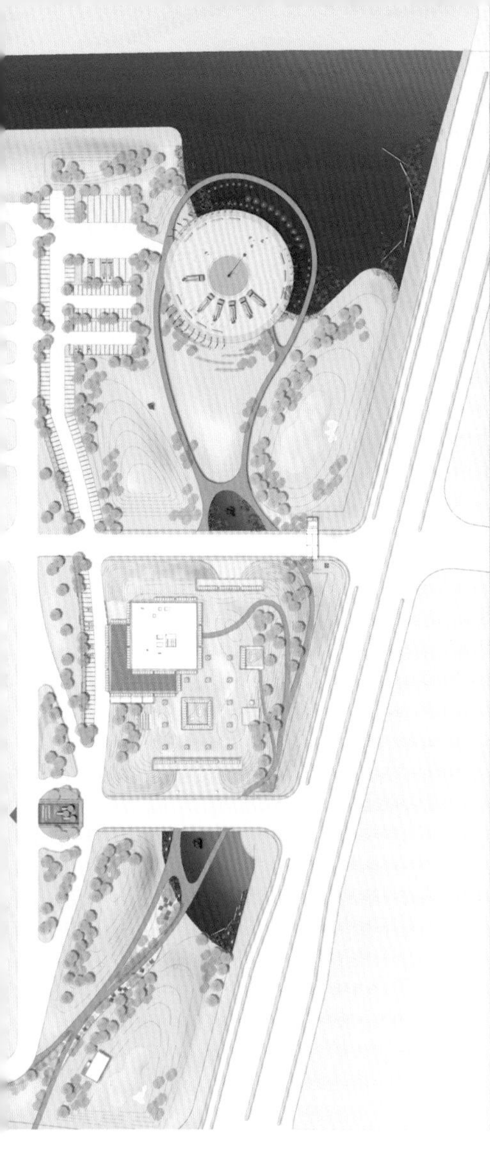

英国伯明翰迎来百年来的第一座新公园

近日，在英国伯明翰，由伦敦的 Patel Taylor 事务所与法国景观事务所 Allain Provost 合作设计了“东边城市公园”(Eastside City Park)，这也成为伯明翰 130 年来第一座新建的公园。

东边城市公园坐落于一个长 800 米的棕地上，向东一直延伸到公园路 (Park Street)，并沿着 Grimshaw 设计的 Millennium Point 设计。其公共娱乐空间有 3.4 公顷，还有 188 米长的运河和 21 座喷泉。

这座直线型的公园还设置了草坪和公共广场等，穿孔的科滕钢 (Corten Steel) 雕塑模仿了旁边的大树造型。

公园耗资 1250 万英镑，成为伯明翰东边复兴区的主要部分，也是伯明翰市议会“大

城计划”(Big City Plan) 市中心再开发计划的一部分。据悉未来 20 年内，再开发计划将耗资 100 亿英镑。

资讯来源：自由建筑报道 http://www.far2000.com/

世界最长蹦床亮相俄景观艺术节

由俄罗斯萨尔托建筑设计事务所设计的世界上最长的蹦床日前亮相 2012 年俄罗斯景观设计国际艺术节。蹦床长 170 英尺（约合 51.8m），相当于 13 辆双层巴士的长度。这张庞大的蹦床看似一条碎石小径，实际却是一条由增强橡胶制成的弹跳跑道，吸引不少参观者在这条蹦床“跑道”上蹦一趟，俨然成为展会明星展品。该艺术节每年在俄南

部尼古拉•莱尼维茨森林举办，被誉为“创意实验室”，创作范围倾向建筑学领域。本次选址森林公园做展地，为艺术家们提供了比城市更大的空间。

资讯来源：中国日报网 http://www.chinadaily.com.cn）

迪拜拟投资 27 亿美元建设五座主题公园

今年以来旅游业的兴旺和地产的复苏坚定了迪拜政府的投资信心。近日，迪拜酋长办公室宣布将投资 100 亿迪拉姆（约合 27 亿美元）建设 5 座主题公园，包括印度宝莱坞主题乐园、海洋公园、儿童乐园和夜间动物园等。此前，迪拜酋长默罕默德还宣布了一项宏大的旅游业和零售业开发计划，包括建设 100 座酒店和世界上最大的购物城。

资讯来源：驻阿联酋经商参处

米其林发布“园林指南”网罗法著名花园景观

据法国《旅游业》网站 10 月 28 日报道，著名的“餐饮圣经”《米其林指南》的打造者宣布发布一款全新的权威评级《法国最美园林指南》，要把法国最著名的花园景观一网打尽。据悉，该系列最新的这本《园林指南》囊括法国 207 个知名园林和 300 余幅精美照片，从国家遗产评定、园林面积、交通状况和服务设施等多种角度来评价各地花园，所有数据经过专家严格审核，并延续著名的“三星打分制”。

资讯来源：环球时报 http://www.huanqiu.com/

英国欲在泰晤士河河口建好莱坞主题公园

据新华社电，法国拉法基水泥公司、英国开发证券公司和加拿大建筑商布鲁克菲尔德集团三家公司牵头共同组新建财团并计划投资 20 亿英镑在泰晤士河河口共同开发建设以美国派拉蒙电影为主题的公园，与知名的巴黎迪斯尼乐园抗衡。该公园占地 872 英亩，位于肯特郡北部。将是英国建立的第一家好莱坞电影主题公园。公园计划在 2018 年运行。

资讯来源：京华时报 http://news.jinghua.cn/

印尼茂物植物园建成尸腐花 DNA 银行

印度尼西亚茂物植物园与印度尼西亚科学院合作建成首个尸腐花 DNA 银行。据茂物植物园负责人慕斯泰德介绍，该园科研人员与印度尼西亚科学院合作，成功培育了 Amorphophalus Titanum 尸腐花。在此基础之上，科研团队利用杂交技术培育出 200 多颗果实，其种子发芽率在稳定在 70% ~ 100% 间。他同时表示，该项研究对于培育其他优良植物具有借鉴作用。

尸腐花，又名“尸臭魔芋”，生长于印度尼西亚。至少 10 年才能开一次花，完全绽放时，花朵直径可达 90 ~ 120cm，花瓣有雨伞那么大。尸腐花吸引人的是它与众不同的“花香”，此花盛开时散发出如尸体腐败般的恶臭，在 0.8km 以外，花的恶臭就能被闻到，而且到了夜晚，臭味会更加浓烈，以便引来帮它传粉的昆虫，故得此名。

资讯来源：科技日报

第三届园冶高峰论坛在北京隆重举行

贺信

住房与城乡建设部副部长 仇保兴

（二〇一三年一月十二日）

尊敬的顾秀莲副委员长：

各位专家学者，同志们：

在全国各地全面贯彻落实党的十八大精神之际，召开这次园冶高峰论坛，很及时也非常有意义，首先我代表住建部向大会的召开表示热烈的祝贺！向获奖的师生、单位表示祝贺！

党的十八大明确提出了以人为本，以科学发展观为引导，大力推进生态文明，建设美丽中国的目标。习近平总书记强调指出，党的十八大把生态文明建设纳入中国特色社会主义事业总体布局，使生态文明建设的战略地位更加明确，这是我们党对社会主义建设规律在实践和认识上不断深化的重要成果。城市园林绿化事业是生态文明建设最有效的手段和重要组成部分。建设美丽家园，需要风景园林师的智慧和创新，推进生态文明，建设美丽中国成为当代风景园林师的神圣责任、历史使命。

生态文明源于对发展的反思，也是对发展的提升。中国自古以来就有世界园林之母的之称，"道法自然"、"宛自天开"、"巧于因借，精在体宜"等理论，都是"园冶"论著的精髓。对此，继承和弘扬园冶艺术，发展和传承园冶文化，也是生态文明建设的重要内容，园冶理论思想对当今风景园林建设和未来风景园林事业的发展仍有启示。希望通过园冶高峰论坛，汇集国内外的专家学者，积极的探索发展，努力实现共同繁荣。

全国建设系统，园林部门要把贯彻落实好加强生态文明，建设美丽中国的伟大号召作为当前和今后工作的出发点和落脚点。要坚持科学发展观为指导，坚持以人为本，坚持生态优先的原则，统筹城乡发展，统筹区域发展，统筹人与自然的和谐，必须树立尊重自然、顺应自然、保护自然的生态文明理念，注重自然和文化资源的保护，开展节约型园林的建设，高度重视和切实抓好园林城市群的建设，园林文化建设、城市生态圈建设和园林产业带建设，着力推进绿色空间、集约发展、文化立意的生态景观空间格局和方式，实现风景园林事业的永续发展。

最后，在新春佳节即将到来之际，祝各位专家学者，行业同仁新年快乐，万事如意！

仇保兴

二〇一三年一月十二日

住房和城乡建设部副部长、党组成员仇保兴发来贺信

开幕式现场

2013 年 1 月 12–13 日第三届园冶高峰论坛在北京新大都饭店成功举办，论坛主题为“生态文明，美丽中国——传承园冶造园文化遗产，建设中华民族美丽家园”。

十届全国人大常委会副委员长顾秀莲莅临大会，会上住房和城乡建设部副部长仇保兴，中国工程院院士、中国花卉园艺与园林绿化行业协会会长金鉴明为第三届园冶高峰论坛发来贺信。

开幕式到会的嘉宾有：十届全国人大常委会副委员长顾秀莲，中国工程院院士尹伟伦；中国工程院院士、中国花卉园艺与园林绿化行业协会副会长卢耀如；文化部华夏文化遗产保护中心理事长乔申乾；中国建设教育协会会长、园冶杯风景园林国际竞赛组委会主任委员李竹成；原住建部城建司副巡视员、中国风景名胜区协会副会长、中国花卉园艺与园林绿化行业协会副会长曹南燕；第九届园博会组委会办公室副主任、北京园林绿化局副局长强健；世界风景园林大师、法国凡尔赛国立高等风景园林学院教授 Henri•Bava；台湾造园景观学会名誉理事长、世界园林杂志总编王小璘；韩国生态景观协会会长、韩国江陵国立大学教授赵泰东；第九届园博会组委会办公室副主任、原北京市政府办公厅副主任陆铭琦、 中国绿色碳汇基金会总工程师李金良，环境保护部、国土资源部、国家林业局以及各省市政府园林局、农业局、园林绿化协会、园艺学会的领导和国内外的风景园林设计机构的负责人，园林企业代表等 600 人。

园冶杯优秀指导教师领奖

获奖企业领奖

第九届园博会设计师广场的获奖设计师领奖

开幕式嘉宾

十届全国人大常委会副委员长顾秀莲

中国工程院院士尹伟伦　中国工程院院士卢耀如

世界风景园林大师、法国凡尔赛国立高等风景园林学院教授 Henri•Bava

中国建设教育协会会长、园冶杯风景园林国际竞赛组委会主任委员李竹成

文化部华夏文化遗产保护中心理事长乔申乾

第九届园博会组委会办公室副主任陆铭琦

中国绿色碳汇基金会总工程师李金良

北京市园林绿化局副巡视员廉国钊

重庆市九龙坡区人民政府副区长周进

国家林业局退耕还林办公室总工敖安强

环境保护部科技司工程师陈永华

作为风景园林行业重要的年度盛会，园冶高峰论坛已经成功举办了两届。本届论坛同期举行了“园冶杯风景园林国际竞赛（毕业设计、论文）”、“园冶杯住宅景观奖”、“城市园林绿化综合竞争力百强暨十佳园林企业”以及“第九届园博会设计师广场”颁奖典礼。

本届论坛持续两天，12 日晚上的风景园林师沙龙邀请到了西安市市容园林局副局长吴雪萍，清华大学景观系副主任朱育帆，台湾造园协会名誉理事长、《世界园林》杂志总编王小璘，翻译马一鸣先生，欧洲著名景观设计师亨利•巴瓦，华中农业大学副校长高翅，易兰国际设计公司副总裁唐艳红（如左下图，左起），与来自国内的众多知名景观设计师进行面对面的交流和探讨，由台湾造园协会名誉理事长、《世界园林》杂志总编王小璘教授主持。

风景园林师沙龙现场

亨利•巴瓦与现场观众热情合影

朱育帆教授与现场观众热烈交流中

园冶杯风景园林国际竞赛（毕业设计、论文）一等奖颁奖

论坛现场

论坛中观众提问

13日举行了多场次的分论坛，围绕“风景园林师的引领与实践”、“当代风景园林师的责任和义务”、“国学造园学术研讨会”以及“社区园林景观的营造”、“园林企业创新管理与企业发展”、“园林绿化行业发展研讨会”等一系列前沿话题进行讲座、对话交流。

主持嘉宾：曹南燕 住建部城建司原副巡视员、中国风景名胜区协会副会长、中国花卉园艺与园林绿化行业协会副会长

台湾造园协会名誉理事长、《世界园林》杂志总编王小璘

第九届园博会组委会办公室副主任、北京园林绿化局副局长强健

华中农业大学园艺林学院院长包满珠

分论坛之——园林城市与人居环境

该分论坛以“生态文明——风景园林师的引领与实践”为主题，由中国风景名胜区协会副会长、中国花卉园艺与园林绿化行业协会副会长（原住建部城建司副巡视员）曹南燕女士主持。会上，台湾造园景观学会名誉理事长、《世界园林》杂志总编王小璘，北京园林绿化局副局长强健，北京公园中心总工程师李炜民，武汉园林局局长苏霓斌，西安市市容园林局副局长吴雪萍，中国园艺学会副理事长、华中农业大学园艺林学院院长包满珠，苏州园林局副局长杨辉，唐山园林局副局长李秀云就生态环境、园林城市建设等问题做了精彩演讲。

主持嘉宾：EDSA Orient 总裁兼首席设计师李建伟

北京清华同衡规划设计研究院副院长、风景园林研究中心主任胡洁

英国AA建筑学院教授、香港大学客座教授、英国普拉斯马公司首席设计师 Eva Castro

北京源树景观规划设计事务所合伙人白祖华

风景园林师论坛暨园林设计院 2012 年会

分论坛之——风景园林师论坛暨园林设计院 2012 年会

该分论坛的主题是“社区园林景观的营造”。EDSA Orient 总裁兼首席设计师李建伟先生担任主持人，会上，英国AA建筑学院教授，香港大学客座教授，英国普拉斯马公司首席设计师 Eva Castro，北京园林古建设计研究院副院长张新宇，棕榈园林股份有限公司棕榈景观规划设计院院长张文英，德国戴水道设计公司 景观设计师总监 Florian Zimmermann（付德景），德国戴水道设计公司景观设计师总监、中国区总监孙峥、西南大学园艺园林学院副院长、西南大学园林景观规划设计研究院副院长张建林，北京源树景观规划设计事务所合伙人、首席设计师白祖华，重庆园林协会副会长、重庆金点园林股份有限公司、重庆高地景观设计公司董事长龙俊，清华大学博士郑晓笛，夏岩文化艺术造园集团研发部总经理吴再悦就人居环境和绿色设施等话题做了精彩演讲。发言嘉宾与参会者以问答方式进行互动，气氛活跃，2012 年年会在热烈的掌声中结束。

主持嘉宾：张树林 中国风景园林学会原副理事长、北京市园林局原副局长

清华大学景观系副主任朱育帆

中南林业科技大学风景园林学院院长沈守云

台湾中华文化大学副教授林开泰

韩国生态景观协会会长、韩国江陵国立大学教授赵泰东

分论坛之——风景园林与科学发展观论坛

该分论坛的主题是“魅力中国－当代风景园林师的责任和义务”，由北京园林学会名誉理事长、中国风景园林学会原副理事长、北京市园林局原副局长张树林主持，韩国生态景观协会会长、韩国江陵国立大学教授赵泰东，华中农业大学副校长高翅，清华大学景观系副主任朱育帆、台湾中华文化大学教授林开泰，中南林业科技大学风景园林学院院长沈守云，清华大学规划设计院风景园林规划研究中心主任胡洁，南京林业大学风景园林学院副院长赵兵，华南农业大学城市规划与风景园林系主任李敏，四川农业大学风景园林学院院长陈其兵，中央美术学院建筑学院景观设计专业负责人丁圆等国内外知名高校园林景观界的专家学者。大家对当代风景园林学科建设以及景观设计师在美丽中国的建设中应发挥怎样的作用等问题发表了自己的看法。现场气氛热烈、专家们的精彩演讲不时博得阵阵掌声。

主持嘉宾：济南园林局副局长刘建东

中外园林建设有限公司董事长王金满

北京市园林古建设计研究院院长张丽华

大千生态景观股份有限公司董事长许忠良

山东大地园林有限公司董事长孙红卫

花王国际建设集团有限公司董理局主席肖国强

分论坛之——园林企业管理论坛

该分论坛主题是“创新管理与企业发展”，由曹南燕会长和济南市城市园林绿化局刘建东副局长主持。论坛上，中外园林建设有限公司董事长王金满、威海园林建设集团工会主席、人力资源部部长刘文元、北京市园林古建设计研究院院长张丽华，苏州园林发展股份有限公司总工程师黄勤、大千生态景观股份有限公司董事长、江苏大千设计院有限公司董事长许忠良、花王国际建设集团有限公司董事局主席肖国强、山东大地园林有限公司董事长孙红卫、江苏山水环境建设集团股份有限公司董事长姚锁平、夏岩文化艺术造园集团常务副总裁苏志勇纷纷分享了公司发展进程中取得的经验和对行业发展的看法。会议结束，曹南燕会长作了总结发言，她希望园林企业健康成长，把美丽中国思想和党的大政方针结合到每个人的工作当中，企业越办越好。

首届国学造园主题论坛

分论坛之——首届国学造园主题论坛

该分论坛由南京师范大学中北学院院长姚海明主持，北京林业大学教授唐学山，第九届园博会指挥部协调处处长崔勇，北京林业大学园林学院教授杨赉丽，西安市容园林局副局长吴雪萍，天津大学建筑学院教授董雅，中国艺术研究院研究员、北京非遗保护工作专家委员会委员孙建君，同济大学美学与艺术批评研究所所长万书元，深圳大学艺术与设计学院院长吴洪，华东理工大学旅游管理系系主任居阅时，中华文化促进会副主席、原文化部文化创意产业司司长王永章，清华大学文化产业研究院教授李季，从造园技法、中国传统园林艺术文化等多方面对国学造园做了深入探讨，并在最后的记者提问环节里，专家们对提出的园林建设与城市建设之间的关系以及中国的年轻一代应该如何传承传统文化等问题一一作了精彩解答。

源树景观（R-land）是国内顶级的专业 环境设计机构。自2004年成立以来，通过不懈的努力，在景观规划、公共空间、旅游度假、主题设计等领域都获得了傲人的成绩，特别是在高端地产景观的咨询及设计方面，处于绝对的领先地位。

源树景观（R-land）的设计团队中汇集了大量的国内外景观设计精英，其主要设计人员都曾在国内外高水平设计单位中担任重要职务，严格的设计流程确保了每一项设计作品的完美呈现。

源树景观（R-land）历经数年，已完成了数百项设计任务，其中：河北省邯郸市赵王城遗址公园、中关村创新园、山东荣成国家湿地、西安大唐不夜城、北京汽车博物馆、龙湖“滟澜山”、天津团泊湖庭院、招商嘉铭珑原、远洋傲北、中建红杉溪谷、西山壹号院等若干项目均已建成并得到各界的广泛认可。

源树景观致力于最高品质的景观营造，力求为合作方提供最高水准的设计保障。

Add：北京市 朝阳区朝外大街怡景园 5-9B（100020） Tel：(86) 10-85626992/3 85625520/30 Fax：(86) 10-85626992/3 85625520/30 - 5555 Http://www.r-land.com

以色列阿瑞尔夏隆公园及海瑞亚填埋场的质变

ARIEL SHARON PARK AND HIRIYA LANDFILL TRANSFORMATION, ISRAEL

彼得·拉茨　　Peter Latz

图 01 2008 年的空中俯瞰图（照片来源：Beracha Foundation）
Fig 01 Aerial view of the site from 2008 （Source: Beracha Foundation）

阿瑞尔夏隆公园排水及景观发展计划
项目位置：以色列特拉维夫
项目面积：840hm²
委托单位：亚肯水利局、阿瑞尔夏隆公园有限公司、贝拉恰耶鲁撒冷基金会
景观设计：拉茨联合事务所
合作单位：水利工程 – 塔豪有限公司、帕齐梅有限公司、MWH
景观设计师 – 布劳多矛兹

海瑞亚填埋场复育
项目位置：以色列
项目面积：118hm²
委托单位：贝拉恰耶鲁撒冷基金会
阿瑞尔夏隆公园有限公司
景观设计：拉茨联合事务所
合作单位：摩利亚撒克里景观建筑师事务所

获　奖
2010 绿色设计竞赛一等奖

Ariel Sharon Park Drainage and Landscape Development Plan
Location: Tel Aviv, Israel
Area: 840 hectares
Client: Yarkon Drainage Authority Park Ariel Sharon Ltd.
Beracha Foundation Jerusalem
Landscape Architects: Latz + Partner
Drainage and Hydraulic: Tahal Ltd with Palgey Maim Ltd and MWH
Local Landscape Architect: Braudo-Maoz

Hiriya Landfill Rehabilitation
Location: Israel
Area: 118 hectares
Client: Beracha Foundation Jerusalem
Park Ariel Sharon Ltd.
Landscape Architects: Latz + Partner
Local Landscape Architect: Moria – Sekely Landscape Architecture Ltd.

Award
Green Good Design 2010 Award

图 02 总平面图－所有层次都被建设成为有所差异变化的景观
（照片来源：Latz + Partner）
Fig 02 The Masterplan–all layers are building up a differentiated landscape
（Source: Latz + Partner）

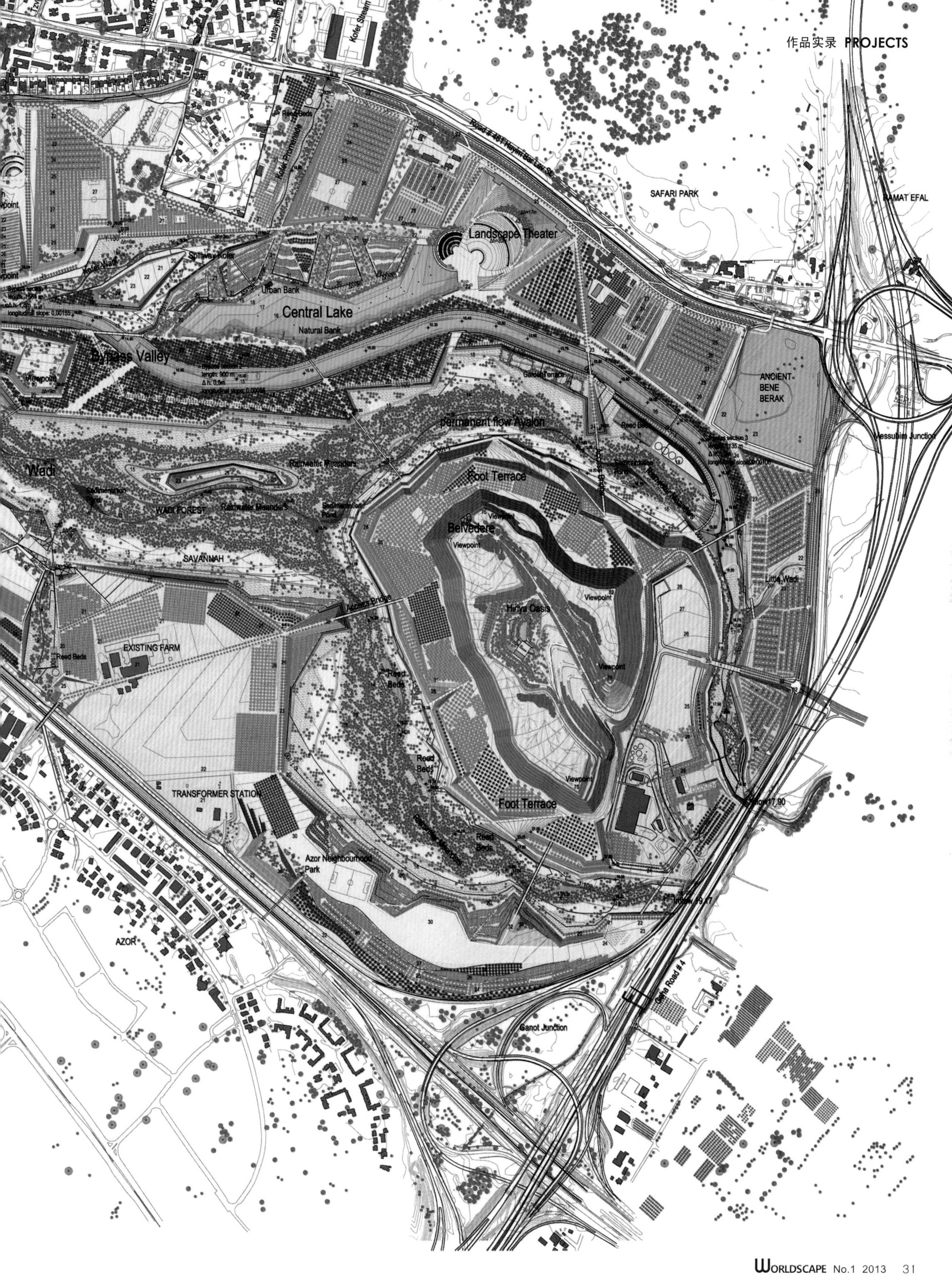

Landscape Theater
Central Lake
Natural Bank
Urban Bank
Bypass Valley
SAFARI PARK
RAMAT EFAL
ANCIENT BENE BERAK
Messubim Junction
permanent flow Ayalon
Wadi
Rainwater Meanders
WADI FOREST
Foot Terrace
Belvedere
Viewpoint
Hiriya Oasis
SAVANNAH
Access Bridge
EXISTING FARM
Reed Beds
Little Wadi
TRANSFORMER STATION
Azor Neighbourhood Park
AZOR
Ganot Junction

图 03 2008 年令人印象深刻的垃圾山坡（照片来源：Beracha Foundation）
Fig 03 The impressive Garbage mountain in 2008（Source: Beracha Foundation）

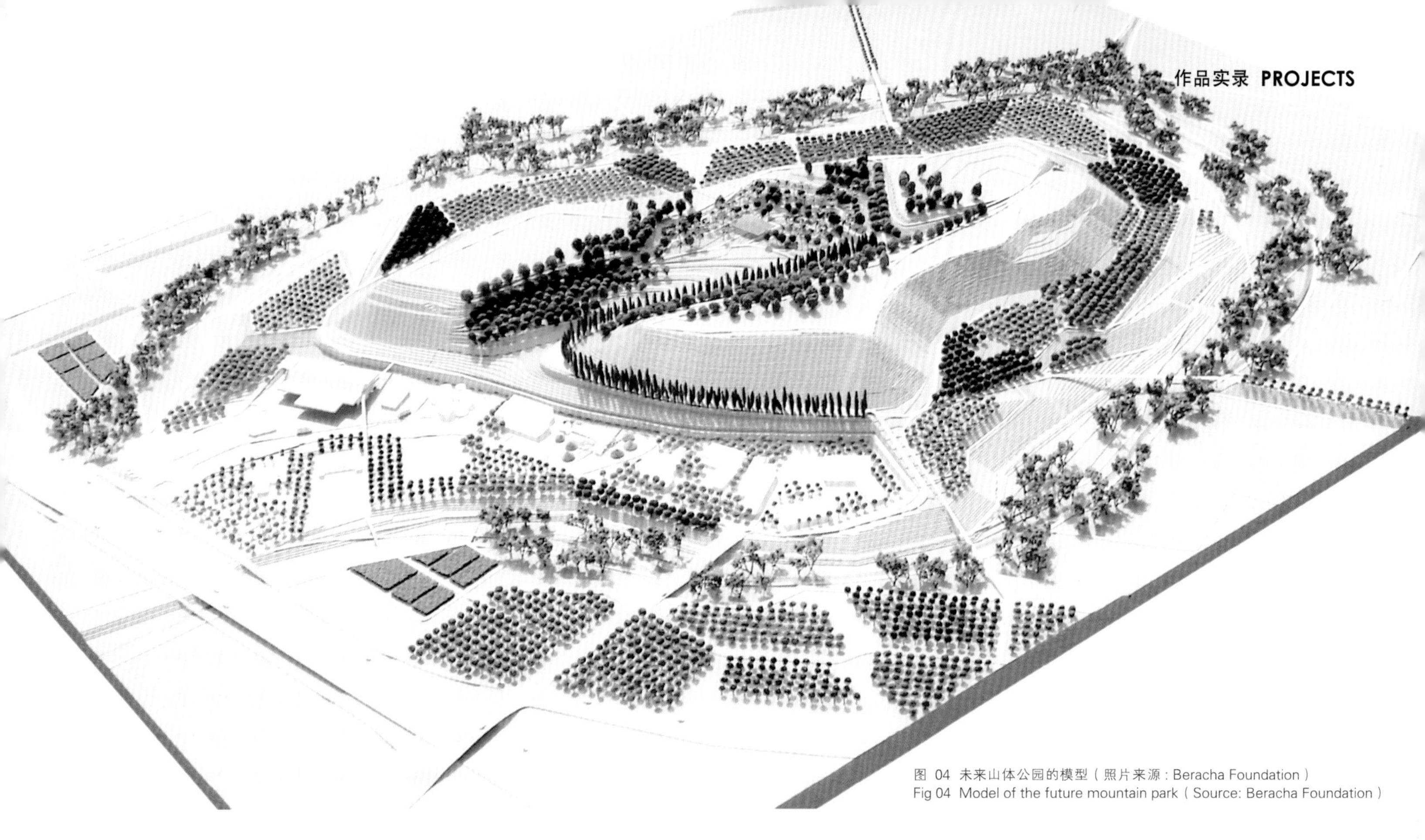

图 04 未来山体公园的模型（照片来源：Beracha Foundation）
Fig 04 Model of the future mountain park (Source: Beracha Foundation)

挑战

所谓21世纪的公园之概念系强调相同场域中科技与自然的结合，以及因此延展出的自明性。然而要将其它元素，诸如在地既存农业之自然性整合进来，显然是不容易的。此一概念是整体将所尝试的核心——与现存概念词汇所展现出的词型变化是完全不同的。'21世纪的公园'也是某种试验，尝试对于城镇与公园关系的重新诠释。将过去一直被避免碰触但却赤裸裸存在的元素整合本身就是个挑战，诸如垃圾堆置场与洪泛保护设施都是（图 01–02）。

公园的任务是必须兼顾生态与教育。但其规划涉及三种不同特性空间场域的堆栈，形成工程上的挑战。分述如下：

1. 海瑞亚，该地区的挑战是确保垃圾填埋场的安全并进行转化（图 03–04）。

2. 阿亚隆平原，其挑战是作为洪水滞留之盆地的服务机能必须被保留下来。

3. 以色列的麦可费，其挑战是保留农业相关设施与功能。

因此，本公园的规划是截然不同过往的。阿亚隆河高水位的特性必须被保留，这是整体规划最重要的部分。到现在，洪水仍不时会侵犯特拉维夫人口稠密地区，所以，关于水患的问题也需考虑。洪水防治计划要求留下可供洪水滞留盆地，并且提供100年防洪频率的洪水72小时的缓冲时间，因此，衍生700万土方的挖掘工程（图 05）。

为避免此巨大的工程冲击邻近的街廓，并防止交通运输穿越住宅区与一般街道，同时考虑遵循生态与永续性的需求，其挖掘出来的土方全数被安置于原基地范围内（图 06）。这些相关数据全都被完整的计算，以确保海瑞亚山垃圾填埋场的安全，这意味着600万立方的再回收弃土，可供挹注于陡峭坡地整理成缓坡的工程中，也因之最新颖的规划理念与以色列麦可费生态导向的农业，得以步入理论发展的阶段。

想象的地景

我们尝试发展可以透过理性途径处理自然的程序。在发展议题上，自然保育区被保留于滞洪盆地中心，我们称其为'河床'（图 07）。洪水径流可以于其间穿越，而预留的河床则可作为大雨或洪泛期间使用。这些水路形成的网络可容纳大量的水运行。

因此，有树荫的自然地景可以进入公园的核心，而这些植物可以自然成长且无需人为的灌溉，甚而形成典型的地中海风格。诸多植生

The challenge

The conception of a park for the 21st century represents the combination, or even the identity, of technology and nature within one and the same space. Moreover, it will certainly not be easy to combine other elements like agriculture - one of the most important existing uses of the area - with the requirement to seem natural. This concept will be one of the biggest tasks -since strong differences exist within the basic paradigms of their conceptual vocabularies.The "Park of the 21st century" is also an experiment which tries to restate its relationship to the town. There is a challenge to integrate totally different elements which were avoided previously, without camouflaging them, such as a garbage dump and flood protection structures (Fig.01-02).

The mission of the park is to be both ecological and educational. It comprises three different spaces characterized primarily by big engineering challenges :

1. Hiriya, with the challenge to secure and to transform a landfill(Fig.03-04),

2. the Ayalon plain with the challenge to serve as a flood retention basin,

3. Mikveh Israel with the challenge to function as an agricultural enterprise.

Quite irrespective of the idea of the park, the retention of high-water of the Ayalon river is one of the most important tasks. Until now, large floods have especially struck populated areas of Tel Aviv, so that they could be considered to be flood disasters. The projected flood protection program requests a retention basin which could be flooded for 72 hours at the most at a 100 years' flood disaster. This results in the excavation of 7 million cubic metres of material (Fig.05).

In order to avoid this gigantic construction project impacting the surrounding urban quarters and to avoid transports through housing areas and on the normal road network and moreover to follow the demands of ecology and sustainability, the excavated

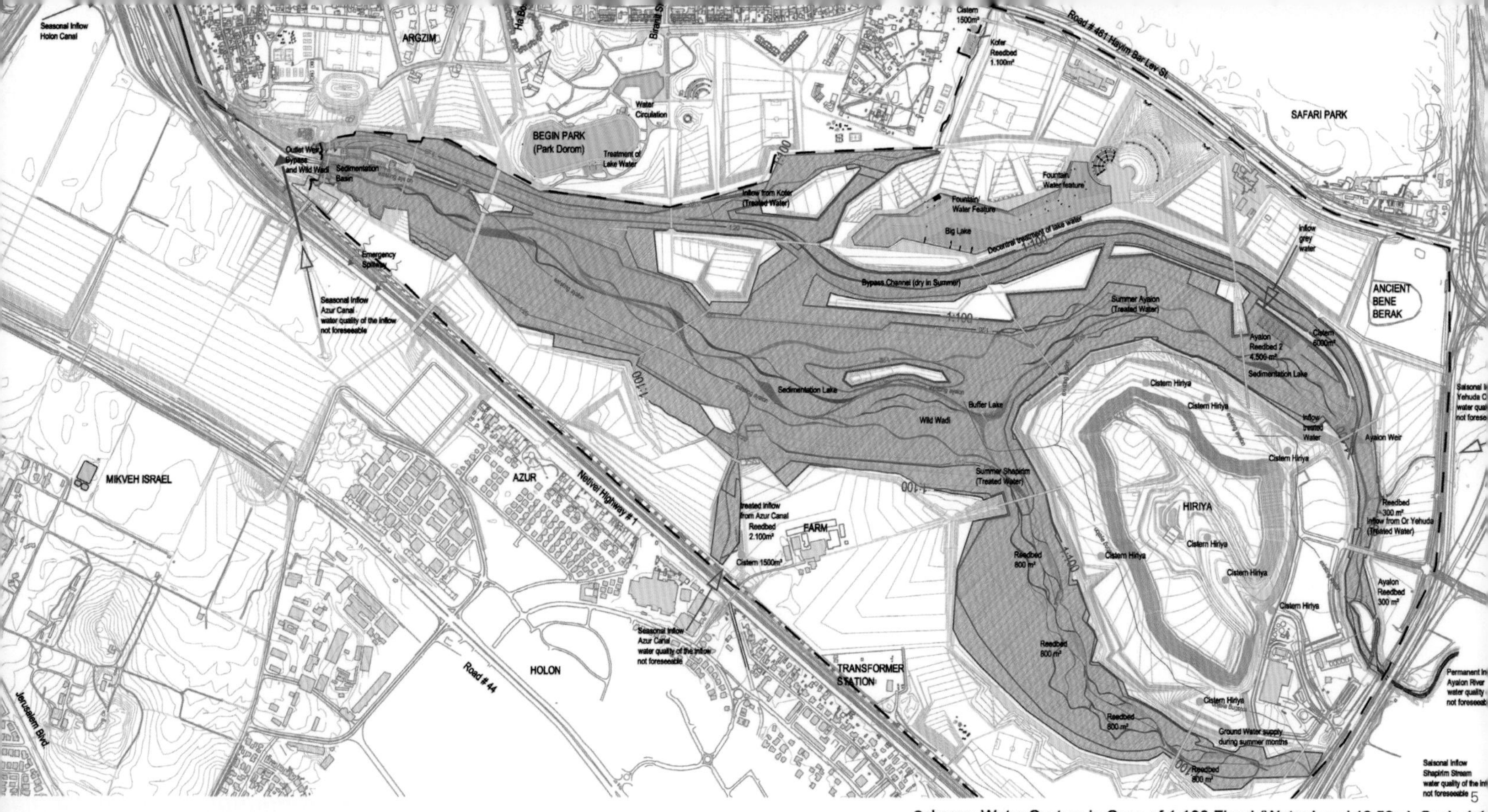

于垃圾填埋场陡峭坡地发展起来，并形成稳定的生态单元，使这些景观解答了后续关于社会性的议题。

人类廊道

首先需要考虑的是人类需求能否与新场域融合，这些场域如何形成，又如何完成，而且形成的根源并非仅仅为邻近居民服务，更在以色列整个国家，甚至国际观光发展所需。这些理念可以通过两种形式的景观予以实现：

1. 首先是“萨瓦讷”，此区域包括公园内最稠密的部份。该地区较‘河床’地势为高，因其稀疏与短急的瀑布使得植物的生长有所不同。座落在未改变与新的建成区，以开凿的人工地景充盈其间，展现戏剧性的张力（图 08)。

2. 其次，最重要的层次在于所谓的‘人类廊道’。多样面向变动的连续带状，沿着位处于公园核心，宛如在处女地的森林荒野间的人行道、场所、水岸区与相关的开放空间跑步（图 09)。这些形成了海瑞亚可支持双脚立足的平台，使沿着运河在新一侧、新的宽阔湖泊与北侧新建的空中剧场慢跑等理想可以成真（图 10)。廊道包含诸多机能和必需的相关公园设施，使它和公园得以全年运作和利用 。

农业纹理

公园的设计应可以从外部被看到，而内部应包含能让人印象深刻的地形地貌景观，毕竟从外部欣赏的人而言，层次感是首先被注意到的。平缓斜坡开凿的露台元素形成了主要地景，而这些地景又被农业的纹理所覆盖；果树与橄榄树构成的林相，与新开凿且为大量绿色植物覆盖的洪水平原接壤；很多的活动可以在植物构成的树荫下产生且彼此融合（图 11）。

可以预测的是，即使作为公园使用的部分已逐渐浮现，现存的农业纹理将可持续很久。然后伴随季节特性进行耕作的方形格状单元、果树与小树林呈现出公园发展的导则（图 12）。

排水计划

与现有地形截然不同的是，我们将在公园内塑造斜坡。为了营造让人印象深刻的地形地貌，我们提供给内向的使用者安静的角落，以及相当多充满风和噪音的场域给外向的使用者，但提供特别的视野景观。地形地貌与植生也提供沿着道路而行时可见的公园‘视窗’，甚至可拓延到海瑞亚山坡地。

我们一方面分析可能作为花园或运动休闲设施的土地使用计划，一方面注意到森林与自然保育区覆盖所衍生的水的问题，甚至基盘设

图 05　科技的复杂系统与空间规划创新，致使公园可以储存达 9 万立方公尺的水（照片来源：Latz + Partner）
Fig 05　A complex system of technical and spacial interventions leads to a park which can store 9mio cubic meters of water（Source: Latz + Partner）

图 06　挖除与回填的方案 – 挖方与填方的平衡创造多样的地形，引导出多元的景色与栖地（照片来源：Latz + Partner）
Fig 06　Sketch/Scheme “Cut and Fill” – the mass balance of excavations and fillings is creating a divers topography hosting a variety of sceneries and habitats.（Source: Latz + Partner）

图 07　经过‘河床’森林紧凑植生小径的慢跑意象（照片来源：Latz + Partner）
Fig 07　Illustration of small paths running through the dense vegetation of the wild Wadi forest（Source: Latz + Partner）

图 08　树木稀疏的平原扮演‘河床’自然地区与人类紧凑廊道的缓冲区（照片来源：Latz + Partner）
Fig 08　The savanna is the transition between the natural area of the Wadi and the desely frequented Human corridor.（Source: Latz + Partner）

masses are to be deposited on the site itself (Fig 06). The fact has also to be taken into account that work is already in process to secure the garbage hill Hiriya. This implies that 6 million cubic metres of recycled demolition waste will support the steep slopes with a terrace at the base of the garbage hill. The idea of up-to-date and ecology-oriented agriculture for Mikveh Israel is still in a theoretical development phase .

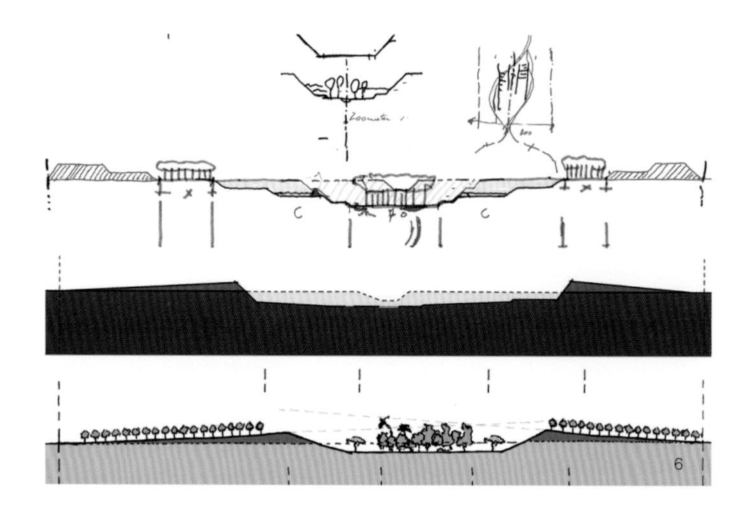

7

8

图 09 人类廊道反应出密集的使用行为，且显露出最频繁出入的区域。如同海瑞亚山的项链般 维护空间，以永续栖地的方式呈现（照片来源：Latz + Partner）

Fig 09 The Human corridor contains all intensive uses of the park and reflects the highly frequented areas. Like a necklace it runs around Hiriya Mountain, keeping space for a sustainable establishment of habitats in between（Source: Latz + Partner）

图 10 中央湖是公园内的主要焦点，且可以由此远眺如同钻石般的海瑞亚山（照片来源：Latz + Partner）

Fig10 The Central lake is one of the main attractions of the park with the view towards the diamond – Hiriya mountain（Source: Latz + Partner）

图 11 农业地也可作为停车使用，游客通常抵达公园，由此向海瑞亚漫步而行（照片来源：Latz + Partner）
Fig11 The agricultural patterns allow also parking. The visitor arrives already in the park, with a view to Hiriya（Source: Latz + Partner）

图 12 山坡与足阶的敷地计划（照片来源：Latz + Partner）
Fig 12 Floor plan of the Mountain and the Foot terrace（Source: Latz + Partner）

图 13 支流的水会被形成古典花园景观的芦苇群所净化（照片来源：Latz + Partner）
Fig 13 The water of the tributary inflows will be purified in reed beds organized in classical garden scenery.（Source: Latz + Partner）

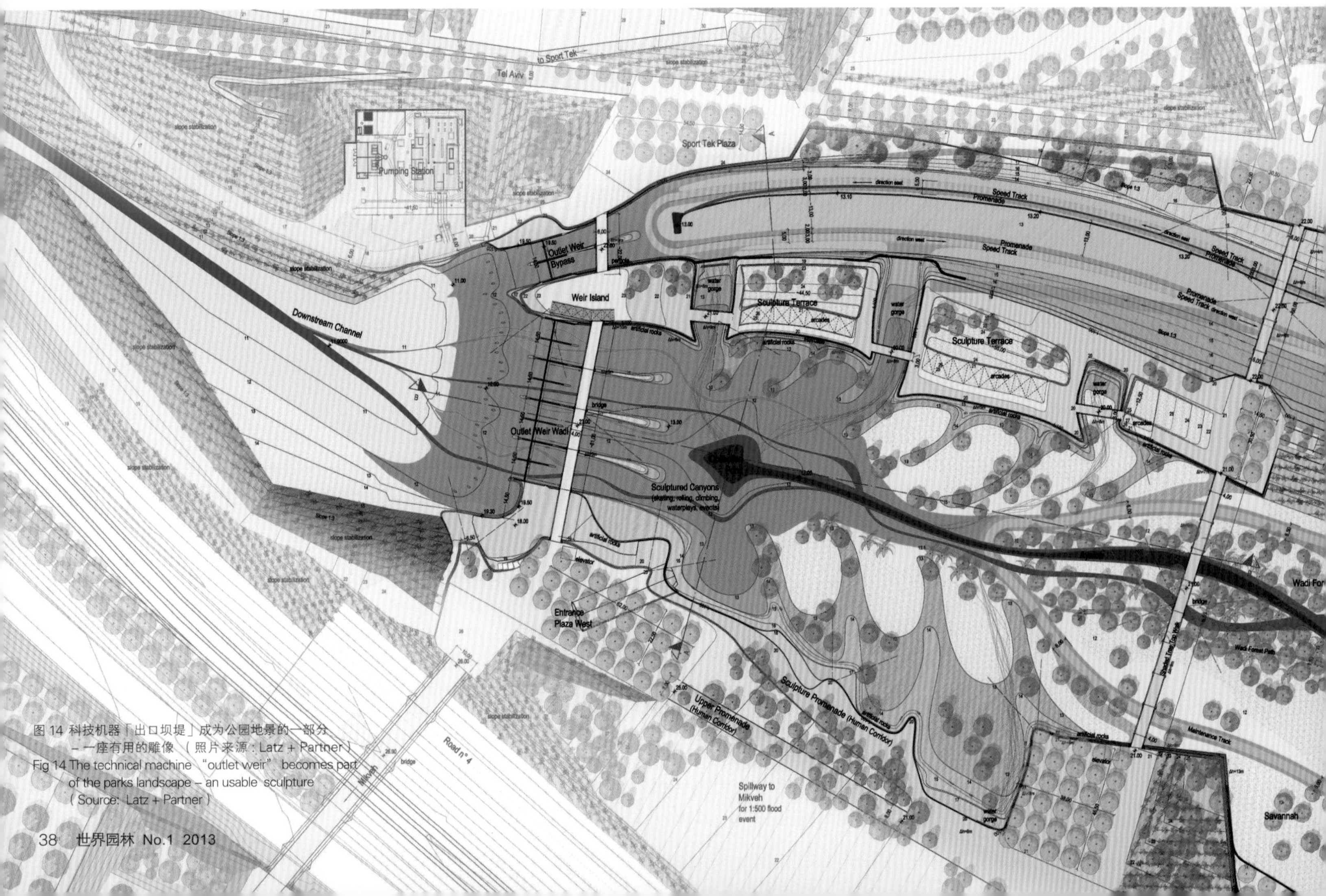

图 14 科技机器「出口坝堤」成为公园地景的一部分——一座有用的雕像（照片来源：Latz + Partner）
Fig 14 The technical machine "outlet weir" becomes part of the parks landscape – an usable sculpture（Source: Latz + Partner）

图 15 在桥上清楚可见入口坝堤引导的水流（照片来源：Latz + Partner）
Fig 15 The inlet weir devides the water streams visible from the bridges around（Source: Latz + Partner）

施、路径系统、街道、公园、服务区域等等，这些都必须考虑其特殊性，以开放供公众使用。

除了关于水资源储存的基本问题外，我们也提出如何清洁水的相关构想。通过群落小区净水的方式（小区坐落于公园中水流经过的途径），水的质量得以提高（图 13）。水的储存池在冬天缺水时挹注，以支持砂石过滤系统的运作，直到夏天月份。另外，对水质的重要保证在于边沟的设计，我们称为'旁道'坝堰，规范了水流且让初期受污染的水体穿越沙石滤层，进入西部边界的混凝土排水系统中；只有干净的、经过中水系统的水体可以被引导到中央储水槽中。因为两河流设计的栏栅可以管控水流进出渠道，中水可以在低水位时被收集起来。栏栅之后的区域提供全面的特殊机能，在夏天时相当有用，且在冬天时成为十分吸引人的水景，可以从桥上看到（图 14–16）。

垃圾填埋场

当下我们对于垃圾或污染区域多觉得眼不见为净，这些区域其特性不容否认，然而过去的产业依然是产业，过去的垃圾掩埋场依然是垃圾掩埋场。但必须将这些特性转换，把它们本来象征的损害与污染，转换成新契机的符号。

海瑞亚仅仅代表整个阿瑞尔夏隆公园的一小部分，但却是相当基础且具意义的指标，马丁威尔与贝拉恰基金会在 1999 年就认知到这一点。目前山的形状具有其独特性，借由陡峭的边坡，我们想要保存可以欣赏这美景的位置，因此，衍生出一些土方工程的需求；举例来说，考虑边坡的稳定性，我们必须从山脚就评估计算力的平衡关系。相关物质的腐败产生的沼气则通过化学程序予以萃取，并收集起来以为他用。

我们的概念引导出后续一连串的行动：如河岸是'河床'的一部分，稳定的足阶、陡峭边坡，高原和内部的绿洲等，皆是在此概念下发展出来的。

从'河床'到足阶间的斜坡，人们可以健行或以单车于和缓的斜坡上行动。穿越'河床'的桥，带出比较高的视野，引领人们导向下个景观元素。我们于此再次建立农业纹理的相关元素，以果树与开花的矮灌丛，去创造出一个由农业使用形塑的景观。这些由路径与小路形成的网络，可以整年被使用，而且人们可以于其间散步、玩耍，而果树林一年中有三季亦可以被充分利用。

第三种元素是陡峭边坡：假如可能的话，我们希望尽可能保留它真实且迷人的形状。此一边坡并非真的死气沉沉，经过时间的淬炼，

Envisioned landscapes

We intend to develop principles which deal with natural processes in an intelligent way. In the course of development, a nature reserve would evolve in the centre of the retention basin: we call it the "Wadi" (Fig.07). Here there will be streams where water flows permanently, whereas other streambeds fill only in case of rain and during the annual flooding periods. A network of paths will allow extensive trips.

Thus, a natural landscape with shady trees could come into being in the core of the park, which after a first phase would be able to grow without artificial irrigation. Thus, a typical Mediterranean formation – the Gariga – should develop on the steep slopes of the garbage hill and perma-cultures could form stable ecological unities. Altogether these landscapes provide answers to the next, the social question .

The Human Corridor

It has to be decided whether and how human needs are met within these new areas; how these spaces could be formed and completed not only for the people living in the neighbourhood but also for the entire state of Israel and international tourism. This can be achieved by two types of landscape:

1. Firstly the "Savannah", which comprises the most densely used part of the park. The savannah lies higher than the "Wadi". As it is dependent on rare and short rainfall, the vegetation will develop differently. Situated between the unchanged and the newly built areas at the rim of the excavation and filling, the artificial landscape promises to become a most dramatic one (Fig.08).

2. So the most important layer becomes the so called "Human Corridor" - continuous belts with changing dimensions, running with promenades, places, water features and other park elements between the virgin-forest-like wild heart of the park and the terraces (Fig.09). They frame the supporting foot terrace of Hiriya, run in parts along the new side canal and concentrate around a new large lake and the planned open air theatre in the north(Fig.10). The corridor contains most of the future functions

图 16 出口坝堰－河床与旁道汇流成地景雕塑
（照片来源：Latz + Partner）
Fig16 Outlet weir – the confluence of the wadi and the by pass as landscape sculpture
(Source: Latz + Partner)

图 17 学生在眺望亭中聆听导览解说（照片来源：Latz + Partner）
Fig 17 A school class listening to the guides explanations in the shade of the Belvedere (Source: Latz + Partner)

沼气污染日渐减少，此地将形成如同‘葛利加’的绿色殿堂，可以发展出热带药草植物，于此植物与美丽的花朵递送着愉悦的气味，可以吸引许多昆虫。

如同陡峭边坡一样，高原也是非常敏感且时时变化而形成其轮廓线。我们思考以边坡低矮及可于贫脊生存之植物延续场域边缘，让这些植物最多生长至膝盖高度，如同此地原生种般自然生长。当植物覆盖后，排水线会形成具有风格的纹理。于此我们发现群山最重要的地方就是“眺望亭”，这是一个观景点，展现特拉维夫像群树般日夜变化的天际线，提供阴影，以穿越阿亚隆平原相当特殊的视野，强调出海瑞亚场域的边界（图 17–19）。

在高原与较深横躺的绿洲，不同的保护层次是必要的：沼气的排放必须被封存，以保护净水渗入山间，水的收集层与覆盖排水系统的植物层。

在下沉区上的内部边缘，高原成了深处最高的台地，此处宛如秘密花园的区域是最重要的元素，形成与外界尺度隔离的自主景观。因此我们称它‘绿洲’，恰好以花园形式融入广阔空间。这些可以和从大高原上收集到的雨水汇合，但也容纳来自山脚下经由沼气设备运作的回收水资源。特殊的灌溉与保水概念，乃透过地下水储存设备，确保植被的茂密（图 20–21）。

不同倾斜度的支撑石墙反映出垃圾的活动，如同表层、过滤层、土壤与排水层，这些都可借由对废弃物的循环再生，获得新生与重新利用的机会。

海瑞亚进行中的植栽绿化是创新且是先驱性的提案。一个新的游客中心提供有关垃圾的信息，也涵盖海瑞亚与公园的发展计划，使民众了解关于垃圾再制的信息，进而了解其新的价值。

面对向环保再生公园方向，我们看到一堵确保边坡安全的墙，游客散步兜风可以到达顶点。扶手镶着金银装饰的棚架，将让人对固着墙壁间的狭小巷弄其平衡性有深刻印象。从露台观看，游客对于整个再生场域可以有完整视野。没有地方隐蔽起来，这将是有趣的经验。□

and necessary park installations. The single sections of these belts are meant to present themselves as complete parks with full equipment and usable all year round.

Agricultural patterns

One should be able to view the park from the outside, therefore the inside contains the expressive topographic landscape, whereas the outside shows almost even levels. The gently sloping terraces built from excavation material constitute a further major landscape element. They are covered with patterns originating from agriculture. Orchards and olive groves follow the edges of the newly excavated flood plain with massive greenery. Under their shady canopies many activities, can be integrated (Fig.11).

It is predictable that the existing agricultural patterns will prevail for a long time even if parts of the park will already be realized. Therefore the rectangular grid with seasonally planted fields, orchards and groves represents one of the guiding principles for the development of the park(Fig.12).

Drainage plan

Totally different from the existing topography, we will have slopes in this park. Due to the expressive topography, there will be “introverted” quiet areas and numerous “extroverted” places which will be windy and often noisy, but offer spectacular views. Topography and vegetation will also offer “windows” along the road to enable views of the park and the mountain Hiriya even from there.

We analysed the possible land use for gardens or for sports and leisure facilities on the one hand and on the other for the realization of water treatment; also of canopies or Macchia, of

图 18 施工中的眺望亭和高原（照片来源：Park Ariel Sharon Company）
Fig 18 The Belvedere and the plateau under construction（Source: Park Ariel Sharon Company）

图 19 夜晚的眺望亭（照片来源：Latz + Partner）
Fig 19 The Belvedere at night（Source: Latz + Partner）

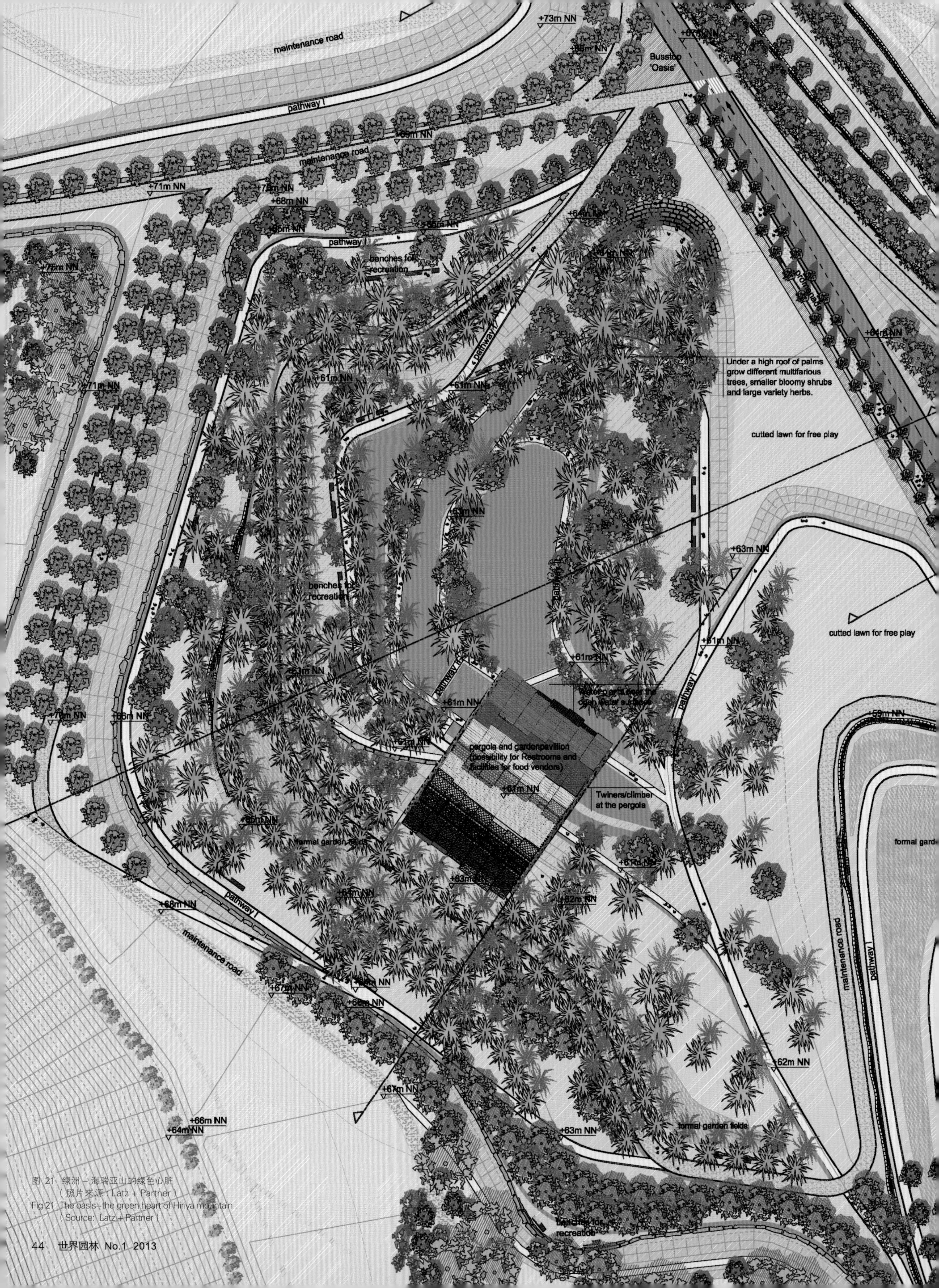

图 21 绿洲－海瑞亚山的绿色心脏
（照片来源：Latz + Partner）
Fig 21 The oasis–the green heart of Hiriya mountain .
（Source: Latz + Partner）

图 20 三层次地景演化出丰富多样的进深，地下水储存池在夏日可提供水的使用。(照片来源 : Latz + Partner)
Fig 20 The three layers of canopies create deep shade,the underground water storage supplies water during summer. (Source: Latz + Partner)

forest and natural reserves; and the infrastructure, path systems, streets, parking, service areas and so on, which is necessary to open the surface of the specific type for use.

Besides the essential question how to store the water, we have to solve the problem of how to clean it. The quality of the water gets enhanced by cleansing biotopes which are installed wherever water flows into the park(Fig.13). Water reservoirs filled in winter secure the water supply for the sand filters during the dry summer months. Another important guarantor for the water quality is a side canal, the so called "bypass". A weir regulates the water inflow and leads the first and usually polluted water mass through the bypass past the Wadi into the direction of the concrete canal at the western edge. Only the clean "second flood" is led to the central retention basin. Due to two river barriers whose outlet is laid out for the maximum capacity of the canal, the second flood collects within the lower lying Wadi. The area behind these barriers will offer spectacular features. They are usable in summer and present a fantastic water spectacle in winter which can be watched from the bridges (Fig.14-16).

Today we cannot afford not to convert wasted and polluted places. These areas should not deny their character and origin: former industry remains industry and a landfill remains a landfill. But they transform their character and features, which stood for damage and pollution, turn into symbols of renewal.

Hiriya represents only a small part of the large Ariel Sharon Park project, but is essential as the iconic starting point identified already 1999 by Martin Weyl and the Beracha Foundation. The present shape of the mountain is unique. We want to preserve this appearance by keeping the steep slopes, bearing in mind the requirements of engineering: for instance the enhancement of the slope's stability with counterweights at the "toe" of the mountain. Methane produced through the chemical processes of decomposition has to be extracted, collected and used for other purposes.

Our concept implies a certain sequence: The riverside, which is part of the 'Wadi', the stabilizing foot terrace, the steep slopes, the plateau and the inner oasis.

On the ramps leading from the "Wadi" to the foot terrace, one can hike and bike easily on gentle slopes. Bridges crossing the "Wadi" on a higher level lead directly to the next landscape element. Here we are also planning to re-establish the element of agricultural patterns, to create a landscape shaped by agricultural use, with fruit trees and blooming meadows. Its network of paths and little roads may be used all year round and one can walk and play on the surfaces in and around the orchards for three-quarters of the year .

The third element is the steep slope: if at all possible it should retain its actual, fascinating shape. The slope is not totally "dead". In the course of time, with the reduction of methane contamination it will green successively like a "Gariga". It will develop a typical Mediterranean vegetation, with plants which flower beautifully and smell pleasantly, attracting many insects.

Just like the steep slope, the plateau is very sensitive to changes in its silhouette. We think that the meagre and low vegetation of the slope should continue over the edge and remain knee-high at the most – like the vegetation which exists here already. Within the plant cover, the green stripes of the drainage lines will form a characteristic pattern. Here we find one of the most important places of the mountain: The "Belvedere" - a viewing point, showing the skyline of Tel Aviv by day and by night. Built like a group of trees, it provides shade and emphasizes the site at the edge of Hiriya with a special view across the wide Ayalon plain (Fig.17-19).

On the plateau and in the deeper lying oasis, different protection layers will be necessary: Sealing off because of the methane emission, which prevents water seeping into the mountain, a water collection layer and a layer of vegetation which covers the layer of drainage.

At its inner edges above the sunken area, the plateau will become the highest terrace of the innermost heart. Inside, the terrace and secret garden are the most important elements, which will form an autonomous landscape without referring in scale to outer spaces. We call it "Oasis", a proper garden within extensively treated spaces; the only place which will get a more intensive shape. It will be irrigated with rainwater collected from the large plateau, but also with recycled water coming from the biogas plant at the eastern foot of the mountain. A special irrigation and retention concept based on underground water storage will enable luxuriant vegetation(Fig.20-21).

The supporting dry stone walls react to the movement of the garbage with different inclines. Like the surfaces, fillings, soils and drainage layers they will be made of recycled construction demolition waste which gets re-manufactured in the recycling plant.

The Hiriya processing plant is an innovative and in the meantime world-famous pilot project. A new visitor centre informs about everything concerning garbage, but also about the plans developed for Hiriya and the park. It informs the public both about the production of garbage and the new values which can be found even in garbage.

Towards the recycling park the slope will be secured by a sheet-pile wall. A visitors' promenade will be built on top of it. A railing and a filigree pergola will create the impression of a catwalk balancing on the retaining wall. From this "balcony", visitors will have great views of the recycling plant. What is usually hidden elsewhere, will here be a focus of interest! ■

作者简介：

彼得·拉茨 / 男 / 工程文凭 / 景观建筑师和城市规划师 / 德国慕尼黑技术大学的荣誉教授

Biography:

Peter Latz / Male / Dipl. Ing. / Landscape Architect and Urban Planner / Prof. emeritus of excellence TU Munich

吴怡彦（中译）何友锋（校译）
Translated by Yiyan Wu, Reviewed by Youfeng He

英国达勒姆遗址海岸－小海岸大讯息

DURHAM HERITAGE COAST, DURHAM, UK—A SMALL COAST WITH A BIG MESSAGE

尼尔·班森

Niall Benson

项目位置：达勒姆，英国
项目面积：530hm^2
委托单位：英国达勒姆县政府
设计单位：达勒姆海岸遗产合伙企业
景观设计：戴维米勒，迈可琼斯，米歇尔麦克伦，约翰古德费洛，艾琳萧
完成时间：2002
获　　奖：2010 年英国园林大奖
2011 年欧洲景观奖理事会特别荣誉奖

Location: Durham, England
Area: 530ha
Client：Durham County Council
Designer: Durham Heritage Coast Partnership
Landscape Design: David Miller, Mike Jones, Michele MacCallam, John Goodfellow, Irene Shaw
Completion: 2002
Awards: The United Kingdom's Landscape Award 2010
The Council of Europe Landscape Awards Special Mention 2011

图 01 飞翔于葵顿的风筝
Fig 01 Kite Flying at Crimdon Credit Mike Smith

背景

达勒姆遗址海岸在 2010 年赢得英国园林大奖，并且在 2011 年代表英国参加欧洲景观奖理事会赢得特别荣誉奖。欧洲景观奖理事会支持“欧洲景观公约”的实施 (注 1)(参 1) (图 01)。

到底达勒姆遗址海岸是如何能得到这份荣耀？背后大有玄机。

不论岩石海岸和沿海草原或是平坦的地板切割的土地和狭窄陡峭的山谷，达勒姆遗址海岸是英国某些最重要的植物群和动物群的天堂，并且还有待挖掘的迷人风光和丰富的历史，达勒姆遗址海岸已成为必访之处。常见各种蝴蝶，如北布朗阿格斯蝴蝶穿梭于点缀于达勒姆岩石盘的草原上，还有一些如蜂兰这种不寻常的兰花，星罗棋布装点出五彩缤纷的夏季草场。游客和当地居民沿岸散步享受风景，此处提供的惊奇美景让来访游客逐年增加。然而，过去并非如此…

20 年前达勒姆遗址海岸还是一片荒地，以黑沙滩臭名远扬，在过去的一个世纪此处为煤炭的倾销地。每年倾倒在达勒姆海滩上的煤渣，巅峰量达 250 万吨。不仅如此，矿坑污水和未经处理的生活污水直接灌注于海洋中 (图 02–03)。

当 10 万人赖以维生的煤矿遭遇关闭的命运，这个国家能源主要生产地迅速转变为社会问题严重和经济贫困的地区。无疑的，此处绝非游客喜爱之处。在葵顿 (Crimdon) 海岸沙滩上，有一个小丽都式度假旅馆，是葵顿 (Crimdon) 小姐年度选拔之处，但这样的设施依旧显得萧条 (图 04)。

规划与回应

为落实矿坑再生议题，投入 1050 英镑以执行“反转波浪”计划

Background

The Durham Heritage Coast won the United Kingdom's Landscape Award in 2010 and went on to represent the UK in the Council of Europe Landscape Awards 2011 and was awarded a Special Mention. The Council of Europe Landscape Awards support the implementation of the European Landscape Convention (Footnote 1) (Ref 1)(Fig.01)

So what is behind such accolades? There is much more to the Durham Coast than fine awards.

From rocky shores and coastal grasslands to woodland denes (deep flat-floored cuts in the land) and gills (narrow steep-sided valleys) the Durham Heritage Coast is a haven for some of the most important flora and fauna in the UK. Coupled with fascinating geology, stunning scenery and a wealth of history waiting to be unlocked and discovered, the Durham Heritage Coast has to be one of the places to visit any year. Butterflies such as the Northern Brown Argus are often seen amongst the rich magnesian limestone grassland that adorns the coast and several unusual species of orchids such as the bee orchid can be seen dotted around the multi-coloured summer meadows. Visitors and locals alike enjoy walking the coastal path throughout the year and an increasing number of people are pleasantly surprised at what this coast has to offer. It was not always like this...

图 02 依辛顿煤矿 1992
Fig 02 Easington Colliery 1992

图 03 依辛顿煤矿 2010
Fig 03 Easington Colliery 2010 Credit: Mike Smith

图 04 1956 葵顿小姐
Fig 04 Miss Crimdon contest 1956

案（1996–2002 年），该计划在 2001 年被认定为遗址海岸而被认同成功。这项目成功促使达勒姆遗址海岸伙伴关系自此成立。他们的工作重点放在促使海岸可及、觉知发展、使用和尊重，以及鼓励游客去探索。

逐渐地，探索者到来，伴随着冰淇淋、咖啡和岩石池的增设。锡厄姆的新码头敞开了大门，今年夏天有 80 个船泊位。相对重要的，尽管与过去功能相去甚远，曾是东达勒姆港煤炭出口的锡厄姆北码头现在有一个新的角色。甚而有之，参观人数大幅增加，2010 年增加 10%，2011 年同比增长 13%（达勒姆郡到访数/STEAM 图/单位通讯），此海岸已规划完成未来 3 年在旅游市场竞争的显著特色。

桑德兰，达勒姆和哈特尔浦沿岸为引导自然英格兰海岸步道计划的先驱（参 2）。两年后这条海岸步道将沿着达勒姆郡桑德兰沿岸直接连到哈特尔浦。新的火车站计划将在 2015 年开放。这些措施都将带

Twenty years ago most of the Durham Heritage Coast was a wasteland, the infamous Black Beaches that had, for a century been the dumping ground for the coal industry. At its peak 2.5 million tonnes of colliery waste was tipped onto the beaches of Durham every year. In addition, minewater and our own untreated sewage was pumped into the sea (Fig.02-03).

The collieries were closed and with this went the livelihoods of 100,000 people who were dependent on the Coal industry. What was once a major energy producing area for the country soon became an area of serious social and economic deprivation. Not surprisingly this coast was not one visitors were likely to visit. There was a small Lido-style resort on the coast's only sand dunes at Crimdon, home to the annual Miss Crimdon competition, but this facility waned as the prospect of Spanish holidays waxed (Fig.04).

Planning and Response

Following the closure of the collieries came the regeneration initiative, the main environmental effort led by the £10.5 million Millennium project 'Turning the Tide' (1996-2002) laid the foundations; its success was marked by gaining Heritage Coast status in 2001. This led to the founding of the Durham Heritage Coast Partnership who have carried the baton ever since. Their work focussing on improving access to the coast, developing awareness, use and respect as well as encouraging visitors to explore.

Slowly the explorers have come, followed by devotees of ice cream, coffee and rock pools. Seaham's new marina opened its gates this summer with 80 berths. Equally important, Seaham's North Dock, once the coal exporting harbour for East Durham now has a new role, a far cry from its origins, but a working port still. There is more to do. Visitor numbers are increasing

图 05 诺斯端点
Fig 05 Noses Point

图 06 燕鸥雕塑
Fig 06 Big little tern Credit Charlie Hedley

来新的观点、新的挑战和新的机遇，并再次为海岸带来变化。

这个经济复苏的故事是真实的，但当地小区是如何看待这事呢？在过去 10 年态度已慢慢改变。对旅游潜力的轻忽态度是根深蒂固的，但对改善当地的环境、信息和游客进入的顽抗态度有较大幅度下降。现在仅有极少数“无此必要”的意见了。伴随着更多的对于环境改善的过程和活动，这些成果和有用的信息已促使当地居民有更积极的氛围。一个曾经激烈反对倾倒垃圾在海岸的本地议员，日前站在一周前开始沿岸割草的承包商的拖拉机和割草机前，算是一个地方“旅游”的小插曲。

越来越多的小区居民使用这多年来不敢造访的的海岸线。前身为矿坑的达顿诺斯端点 (Dawdon Nose’ s Point)，已被达勒姆郡议会改造成一个“门户”，变更为一个进行思考和沉思的安静地点。这为场地管理和确保社会和环境改善的长期安全，带来新的议题和责任，特别是经济恶化时期的财务挑战 (图 05)。

霍登，因工业倒闭受创最重的小区之一，由于穆尔设计的 David Buurma 和当地学校与小区团体而令人鼓舞。童言童语可打动冰冷如钢铁的心，所以曾经屡见不鲜的任意倾倒、车辆横冲直撞和弃置光缆的地方，如今成为家庭漫步之处。国际基金会由每周日早上沿步道默默捡拾垃圾的志工组成，这样的奉献精神令人感到由衷的赞赏 (图 06)。

从自然保育的角度来看，煤炭石并不全是坏事，因为它成为海岸斜坡上独特的莱姆石草原的柔性防御，这令人惊异的栖地被侵蚀到沿岸边缘，环绕着陡峭岌岌可危的海岸斜坡。将自然的草原还原为可耕种的土地的计划已进行逾十年，独特的莱姆石草原正被海洋侵蚀而急遽消失。当务之急，为确保未来这独特草原的存在，更直接的干预可能是必要的 (图 07)。

民众参与

海洋环境又是如何呢？ 1992 年 (参 3)，一份初始的研究调查有

substantially, up 10% in 2010 and 13% in 2011 (Visit County Durham, STEAM figures, pers comm). The coast will feature substantially in the tourism marketing campaign that is being planned for the next three years.

The Sunderland, Durham and Hartlepool stretch of coast features as one of the lead areas in Natural England's Coastal Trail programme (Ref 2). This coastal trail will be delivered locally over the next two years, directly linking Sunderland along the coast of County Durham to Hartlepool. We also have a new railway station being planned to open in 2015. Each of these initiatives will bring new views, new challenges and new opportunities which will bring change to our coast again.

There is real power in the story of this recovery, but how is this viewed by the local community? Slowly attitudes have changed over the past ten years. Local cynicism about the potential of tourism is deep seated. Local cynicism about improving their local environment, information and access has decreased substantially. There are very few ‘there’s no point’ comments nowadays. With much more informed and helpful information being provided and real involvement in activities and design processes the local mood is so much more positive. One local councillor, once strident in defence of vehicular access to the beach which led to severe fly tipping issues, recently stood in front of the tractor and mower of a contractor who had started mowing a coastal hay meadow one week early; an illustration of the local ‘journey’.

Local communities are increasingly using a coastline that they were deterred from using for so long. Nose’s Point, Dawdon, a

图 07 遭受威胁的海岸草原
Fig 07 Coastal grassland under threat

益于想象，其中标注“仅见少量的海洋生物”，并感谢潜水员志工潜入“他们未曾想前往之处”。相对的，2009 年的研究调查 (参 4) 中记录了 94 种物种，显示海洋生物正强力复苏。借由大海调查遗址彩券的资助 (参 5)，当地居民被招募和培训，成为与这片海岸关系密切且最重要的记录员，以提供改善的证据 (图 08)。

参与是这些惊人成就的关键，具直接关连的当地小区提供持续存在的伙伴关系来改善海岸。我们鼓励他们落实实际的计划案，成为超越地区、更宽广的网络。这种伙伴关系为地主、管理者、警务人员和街区督导员提供了一个有用的链接。

结语

遗址海岸成为这类工作的关键。在英国，并没有法定上的规划定义，这意味着它是经由自愿的认可并且没有官方认定的规划方式。因是自愿的，就具有认同它的额外力量，如果没有这力量，将缺乏明确的目标或外界的认可。在英国的“遗产海岸”可以被当成是缺乏国家支持，失落于国家公园和自然风景区间，疏于保护的六零年代景观家族成员。这是一个真实的案例，探讨如何以崭新手法达成满足 21 世纪需求的海岸质量。确保政治认可、实时的财政支持及联结人群与海岸的明确目的，对于达勒姆海岸这片被掠夺之地都将是异常重要。

目前的挑战是如何利用已完成的资源，将此遗址海岸当成温馨、热情的入口门户，向外连接更宽广的海岸和休闲环境，以达到最大可增的游客人数。□

注 1: 欧洲景观公约“ – 也被称为佛罗伦萨公约” – 促进欧洲景观的保护、管理和规划，并组织欧洲景观议题的合作。2000 年 10 月 20 日在意大利佛罗伦萨通过该公约，并且在 2004 年 3 月 1 日生效 (欧洲委员会条约汇编“第 176 号) 。它开放供欧洲理事会成员国、欧洲共同体和欧洲非成员国加入签署，这是第一个专门关注所有欧洲景观面向的国际条约。

参考文献

1.Council of Europe European Landscape Convention web pages: http://www.coe.int/t/dg4/cultureheritage/heritage/landscape/default_en.asp
2.Natural England Coastal Access web pages - http://www.naturalengland.org.uk/ourwork/access/coastalaccess/durham/default.aspx]
3.Loretto, C-J. 1992. Seasearch survey of the Durham coastline. (Contractor: Marine Biological Consultatns). JNCC Committee Report 29.
4.Lightfoot P (2009), Seasearch North East,. Durham Heritage Coast Seasearch Survey Report
5.Big Sea Survey web page: http://www.bigseasurvey.co.uk/

former colliery site that has been transformed into a 'gateway' to the coast by Durham County Council, is being adopted as a quiet site, for reflection and contemplation. This brings with it new issues and new responsibilities for site management and long term strategy to secure the social and environmental improvements for the long term, particularly with the financial challenges of these straitened times (Fig.05).

In Horden, one of the communities hardest hit by the industrial closures, a circular easy access path with the addition of inspirational interpretation has drawn villagers out to the coast, some for the first time. With strong support from the National Trust, interpretation was developed with Mor Design's David Buurma and the participation of local schools and community groups. The children's words cut into the steel panels have the power to move even the hardest of hearts, so that where once fly tipping, car burnouts and cable stripping were commonplace, family groups wander freely. The National Trust are supported by the work of local volunteers, one of whom early every Sunday walks the path and picks off any litter, quietly and unobtrusively; but such devotion is heartily appreciated nonetheless(Fig.06).

From a nature conservation point of view the coal spoil was not all bad as it provided a form of soft defence for the unique paramaritime magnesian limestone grassland on the coastal slopes. Coastal squeeze had reduced this amazing habitat to the very fringes, clinging to the coastal slope like a monastic tonsure, in real and present danger of liquidation. A programme to revert the immediate arable hinterland to more natural grassland has been underway for over a decade. However, as the spoil on the beaches erodes, this unique limestone grassland is rapidly being lost to the sea. The pressure is now on to ensure its survival into the future – more direct intervention may be needed (Fig.07).

Public Participation

How has the marine environment fared? An initial Seasearch survey in 1992 [Ref 3] makes salutary reading, noting 'little marine life was encountered' in the near shore, and thanked the volunteer divers for "diving where they never thought they would". Contrast this with the Seasearch survey in 2009 [Ref 4] where 94 species were recorded, illustrating a strong revival of marine life. With support from the Heritage Lottery Funded Big Sea Survey [Ref 5] local people are being recruited and trained as recorders and to become involved in their local shore and most importantly; providing evidence of improvement (Fig.08).

Participation is key to these amazing achievements; the partnership's persistent presence has provided the local community with a direct connection to an improving coast. We are encouraging them to become connected real physical projects as well as becoming part of a wider coastal network beyond their locality. The partnership also provides a useful link to the landowners and managers, the police force and the street wardens.

Conclusion

Heritage Coast status has been key to most of this work. In the UK this status is a non-statutory planning definition, which means that it is voluntarily recognised and has no legal support in formal planning terms. Being voluntary it has an additional strength in that those who want it will recognise it, without this there would have been no clear vision or external recognition. 'Heritage Coast status' in the UK can be seen as a bit of a sixties curio, lost

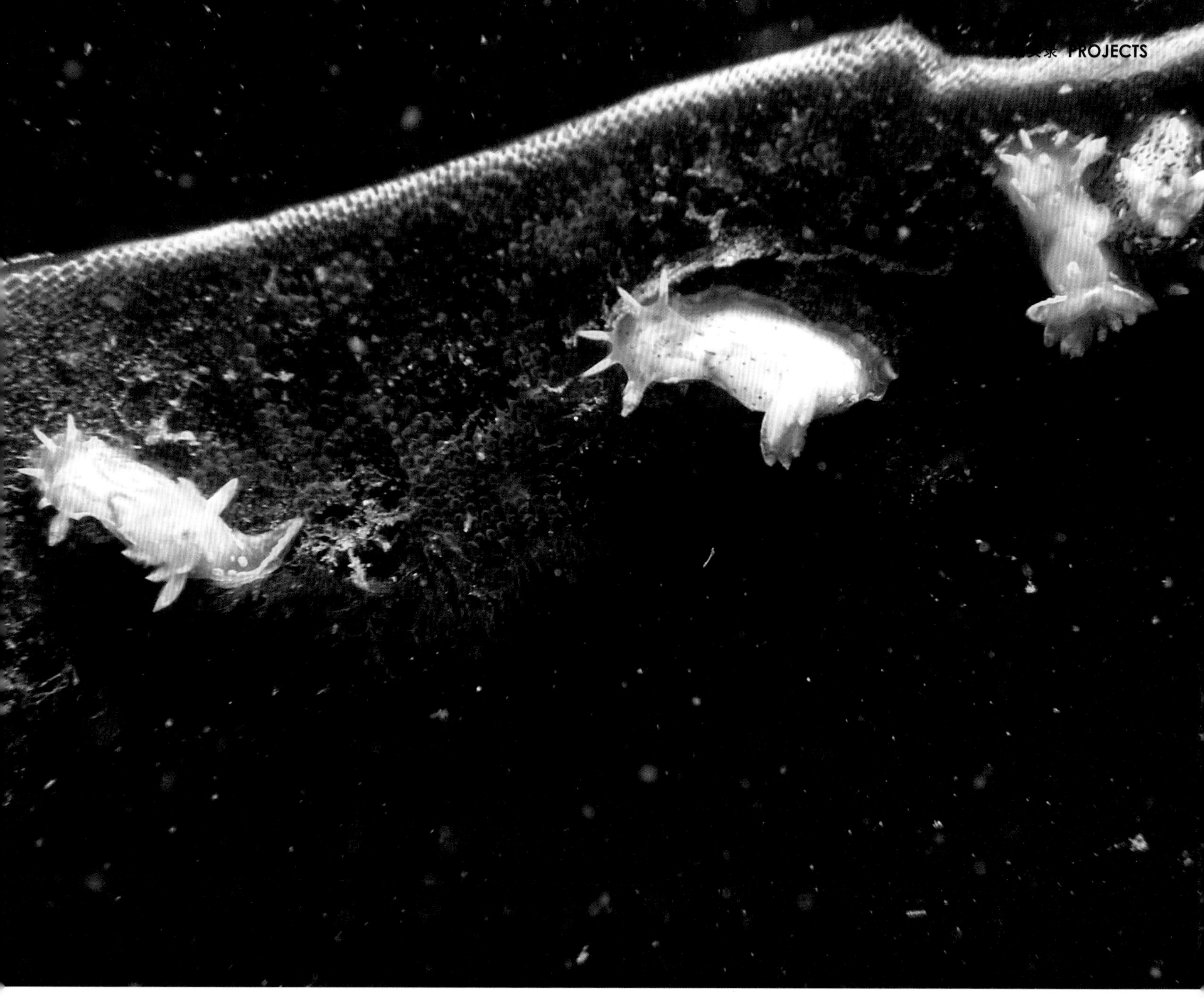

图 08 海带上的裸鳃亚目海洋生物
Fig 08 Nudibranch Polycera quadrilineata on a strand of kelp Credit Paula Lightfoot

amongst the Protected Landscape family, that includes National Parks and Areas of Outstanding Natural Beauty, and has no clear national support. There is a real case to explore how the status can be refreshed to assist in delivering quality coasts for 21st century needs. For the despoiled coast of Durham it has been terrifically important in securing political 'buy-in', critical financial support and that clear purpose that links people to their coast.

The challenge now is to capitalise on what has been done, make the most of the increase in visitor numbers by linking the whole coast through providing warm, welcoming entry points and linking outwards to our wider heritage and recreational environment. ■

Footnote 1: The European Landscape Convention - also known as the Florence Convention, - promotes the protection, management and planning of European landscapes and organises European co-operation on landscape issues. The convention was adopted on 20 October 2000 in Florence (Italy) and came into force on 1 March 2004 (Council of Europe Treaty Series no.176).It is open for signature by member states of the Council of Europe and for accession by the European Community and European non-member states. It is the first international treaty to be exclusively concerned with all dimensions of European landscape

References

1.Council of Europe European Landscape Convention web pages: http://www.coe.int/t/dg4/cultureheritage/heritage/landscape/default_en.asp
2.Natural England Coastal Access web pages - http://www.naturalengland.org.uk/ourwork/access/coastalaccess/durham/default.aspx]
3.Loretto, C-J. 1992. Seasearch survey of the Durham coastline. (Contractor: Marine Biological Consultatns). JNCC Committee Report 29.
4.Lightfoot P (2009). Seasearch North East,. Durham Heritage Coast Seasearch Survey Report
5.Big Sea Survey web page: http://www.bigseasurvey.co.uk/

作者简介：
尼尔·班森 / 男 / 遗址海岸官员 / 达勒姆遗址海岸合伙企业

Biography:
Niall Benson / Male / Heritage Coast Officer / Durham Heritage Coast Partnership

刘正智（中译）何友锋（校译）
Translated by Zhengzhi Liu, Reviewed by Youfeng He

上海辰山植物园矿坑花园

SHANGHAI CHENSHAN BOTANICAL GARDEN, QUARRY GARDEN, CHINA

翟薇薇

Weiwei Zhai

Water Tower

Steep Ladder

Rusty-steel Shutters

Winding Path

Lake Aera

图 01 鸟瞰效果图
Fig 01 Rendering Effect of Aerial View
注：所有图片由朱育帆提供
Copyright: all images provided by Yufan Zhu.

项目位置：上海市松江区
项目面积：4.3hm^2
委托单位：上海辰山植物园
设计单位：北京清华同衡规划设计研究院有限公司
主 持 人：朱育帆
设计团队：朱育帆、姚玉君、孟凡玉、王丹、张振威、冯纾苨、孟瑶、孙天正、严志国、翟微微、郭畅、孙建宇、齐羚、杨展展、崔庆伟、张隽岑、龚沁春、常钰琳、田锦、董顺方、孙珊
完成时间：2010
获　　奖：2010 年 12 月 第二届中国建筑传媒奖最佳建筑奖提名
2011 年英国风景园林行业协会国际门类奖
2012 年美国景观设计师协会综合设计类荣誉奖

Location: Songjiang District, Shanghai, China
Site Area: 4.3 hectares
Client: Shanghai Chenshan Botanical Garden
Design Organization: Beijing Tsinghua Urban Planning & Design Institute
Director: Zhu Yufan
Design Team: Zhu Yufan, Yao Yujun, Meng Fanyu, Wang Dan, Zhang Zhenwei, Feng Shuni, Meng Yao, Sun Tianzheng, Yan Zhiguo, Zhai Weiwei, Guo Chang, Sun Jianyu, Qi Ling, Yang Zhanzhan, Cui Qingwei, Zhang Juncen, Gong Qinchun, Chang Yulin, Tian Jin, Dong Shunfang, Sun Shan
Completion: 2010
Awards: The 2rd China Architecture Media Award Nominee Dec. 2010
2011 BALI National Landscape Awards
2012 ASLA honor award in general design

图 02 从山上鸟瞰全园
Fig 02 Bird’s-eye view of the Garden from the hill.

矿坑花园位于上海辰山植物园西北角，通过绿环道路和辰山河边主路与整个植物园相连。在20世纪以来采石，南坡半座山头已被削去。为保护矿山遗迹，加快生态矿山、美化环境建设，结合上海辰山植物园的建设，这里被批准建设成为一个精致的、有特色的修复式花园。通过对现有深潭、坑体、迹地及山崖的改造，形成以个别园景树、低矮灌木和宿根植物为主要造景材料，构造景色精美、色彩丰富、季相分明的沉床式花园。

作为一项历史悠久的工业活动，采石业伴随着漫长的文明进程而发展，见证了人类活动对自然的干扰、掠夺和破坏。采石工业剥离表层植被，剧烈改变地形，造成水土流失、景观破坏和生境破碎化，是生态退化研究中的一种重要类型。矿坑花园总体面积为4.3hm^{2}左右（图01–02），由高度不同的四层级构成：山体、台地、平台、深潭。其中，山体表面较平整无层次且风化相对严重，无明显纹理和凸凹，无裂纹，立面有直开的矩形通风口，显突兀。台地上植被茂盛，靠近岩壁的位置现状留有洞库的出入口6个；平台部分为采石留下的断面，地势较平，边缘地区有生长良好的水杉林；深潭面积在1hm^{2}左右，与平台层高差约52m，潭水清澈，有自然形成的岛屿和植被带来一丝生机（图03）。

这样一个项目中，设计师面临很多挑战。第一个挑战是修复严重退化的生态环境。场地内植被稀少，物种贫乏，岩石风化、水土流失严重。第二个挑战是充分挖掘和有效利用矿坑遗址的景观价值。因此，如何重新建立矿坑和人们之间的恰当联系成为设计师需要思考的问题（图04）。

设计师选择了同时用“加减法”应对采石矿坑特殊形态的生态修复设计原则：采取“加法”策略通过地形重塑和增加植被来构建新的生物群落。针对裸露的山体崖壁，设计师没有采取常规的包裹方法，而是尊重崖壁景观的真实性。在出于安全考虑的有效避让前提下，设计师采取了不加干预的“减法”策略，使崖壁在雨水、阳光等自然条件下进行自我修复。对于存留的台地边缘挡土墙，设计师用锈钢板这种带有工业印迹的材料，对其进行包裹，形成有节奏变化和光影韵律的景观界面（图05–13）。

在中国山水画和古典文学的审美启示下，该项目采取现代设计手法重新诠释了东方自然山水文化以及中国的乌托邦思想。不同于西方“静观”的欣赏方式，东方传统更强调可观、可游的“进入”式山水体验。设计师在平台去设置一处“镜湖”，倒映山体优美的曲线，从四周都可以观看，增大了观景视域。为了改造山体稍显枯燥的立面，倚山而建一个水塔，有效地调整了其节奏，并有泉水从山中流出，增加生趣。

Quarry Garden lies in the northwestern corner of Shanghai Chenshan Botanical Garden and is connected with the entire Botanical Garden by green circle road and the main road by Chenshan River. Half of the hilltop on the south-facing slope has been chipped off due to quarrying throughout the 20th century. A delicate and distinctive reclaimed garden was approved to be built at this site, in combination with the construction of the Shanghai Chenshan Botanical Garden, with a view to protecting mine relics, speeding up the pace of ecological mine reclamation and beautifying environmental construction. Efforts are made to build a sedimentary mineral deposit garden, featuring individual specimen trees, shrubs and perennial plants as the main landscaping materials as well as a delicate, colorful and seasonal landscape, through the renovation for the existing pool, pit, slash and cliff.

As part of the long march towards progress, the industrial activity of the quarrying industry has resulted in a great deal of damage to nature. The industrial practices of the quarrying industry were an important factor considered in the research

入口 ▲矿坑花园入口 ▲保留出入口
分区 A 深潭区 B 镜湖区 C 台地区
景点 1 花床 2 弧形坡道 3 云梯 4 锈钢板百叶墙 5 锈钢板墙 6 “洞口春云”水塔 7 秘园 8 “甘白山泉” 9 山瀑 10 瀑布 11 锈钢板栈道 12 “丹井灵源”钢筒 13 浮桥 14 望花台 15 现状水杉林 16 一线天 17 通东矿坑隧道

图 03 总平面图
Fig 03 Master plan

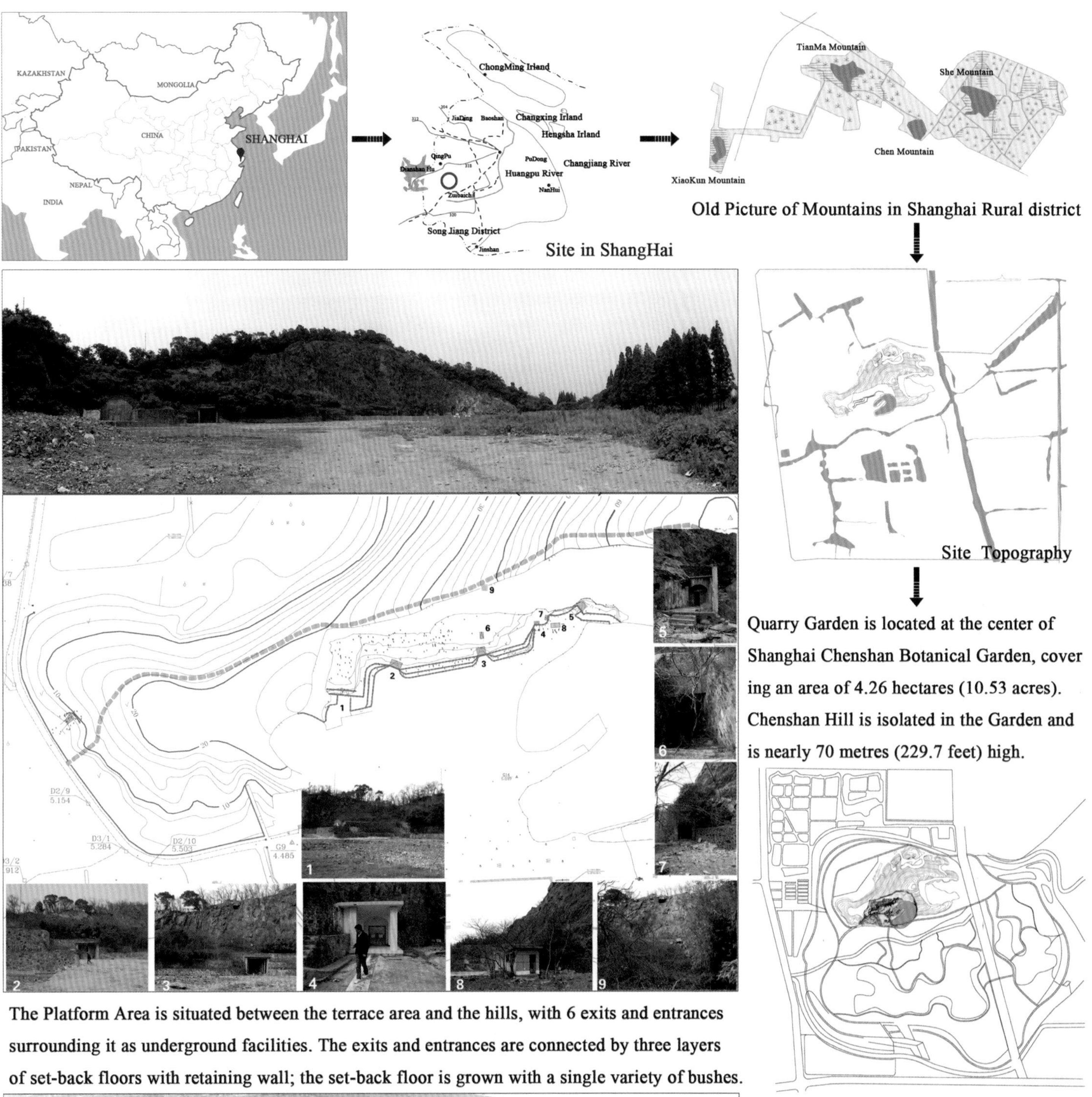

Quarry Garden is located at the center of Shanghai Chenshan Botanical Garden, covering an area of 4.26 hectares (10.53 acres). Chenshan Hill is isolated in the Garden and is nearly 70 metres (229.7 feet) high.

The Platform Area is situated between the terrace area and the hills, with 6 exits and entrances surrounding it as underground facilities. The exits and entrances are connected by three layers of set-back floors with retaining wall; the set-back floor is grown with a single variety of bushes.

The deep pool has an area of about 1 hectare (2.47 acres); both the water depth and the height difference between the water surface and the terrace area is between 20-30 m (65.6-98.4 feet).Due to its unique space form, this area is doomed to make the core zone of this project. Since the deep pool has been affected by the rampway for transporting stones, the rock wall on its south side is much more abundant in space levels.

055-01 Site historical context

图 04 改造前的现场实景图
Fig 04 Site situation

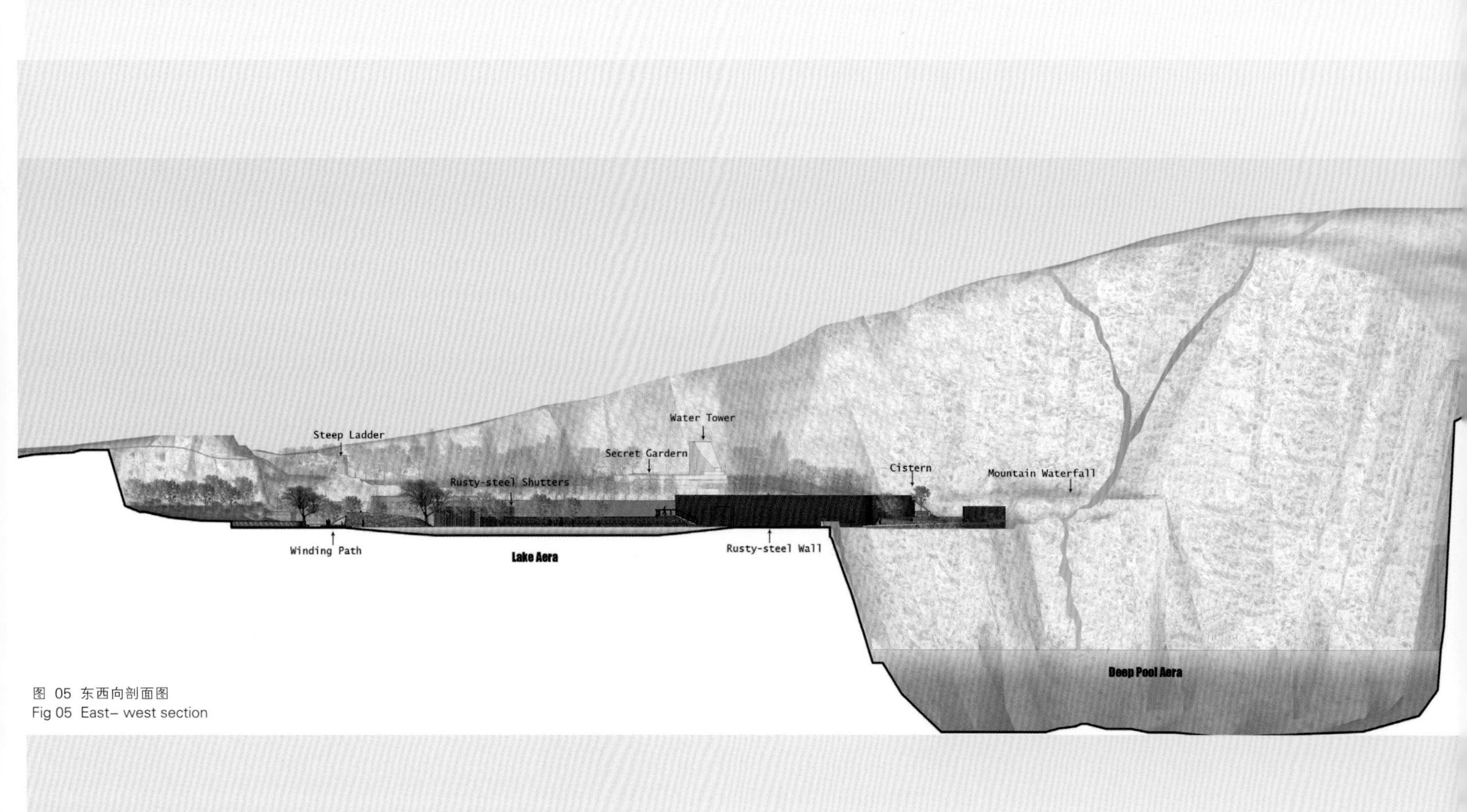

图 05 东西向剖面图
Fig 05 East– west section

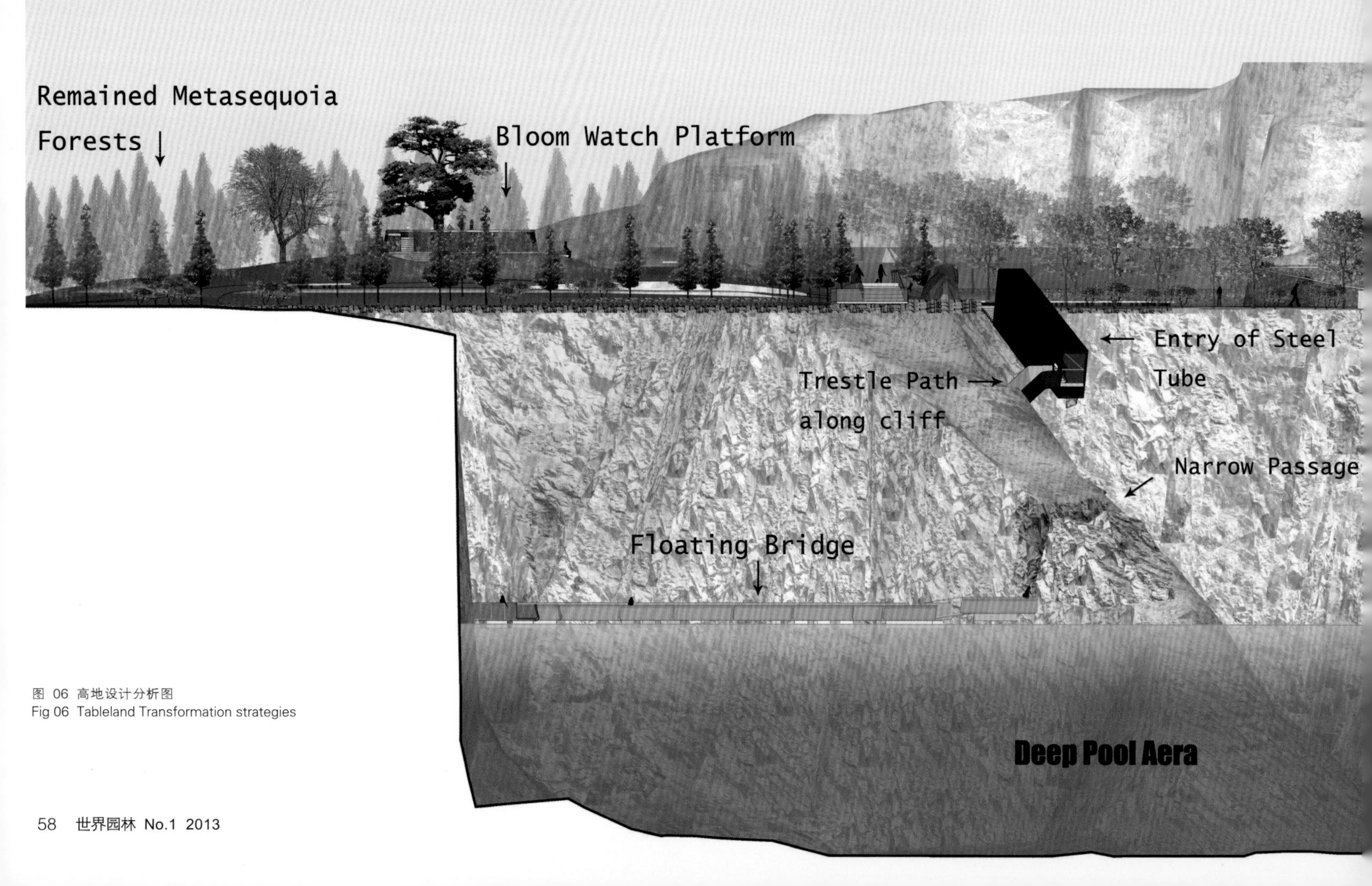

图 06 高地设计分析图
Fig 06 Tableland Transformation strategies

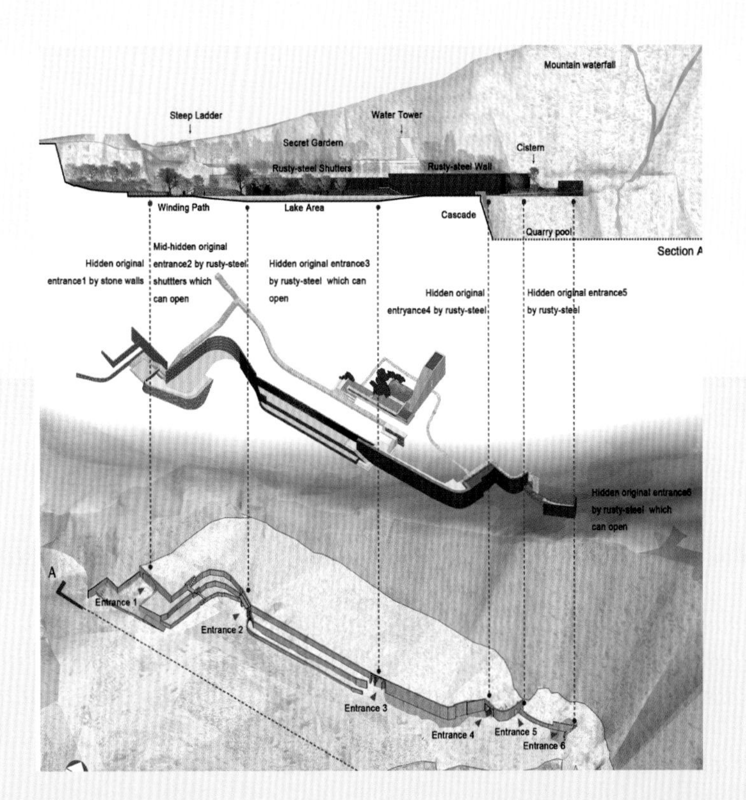

图 07 南北向剖面图
Fig 07 North-south section

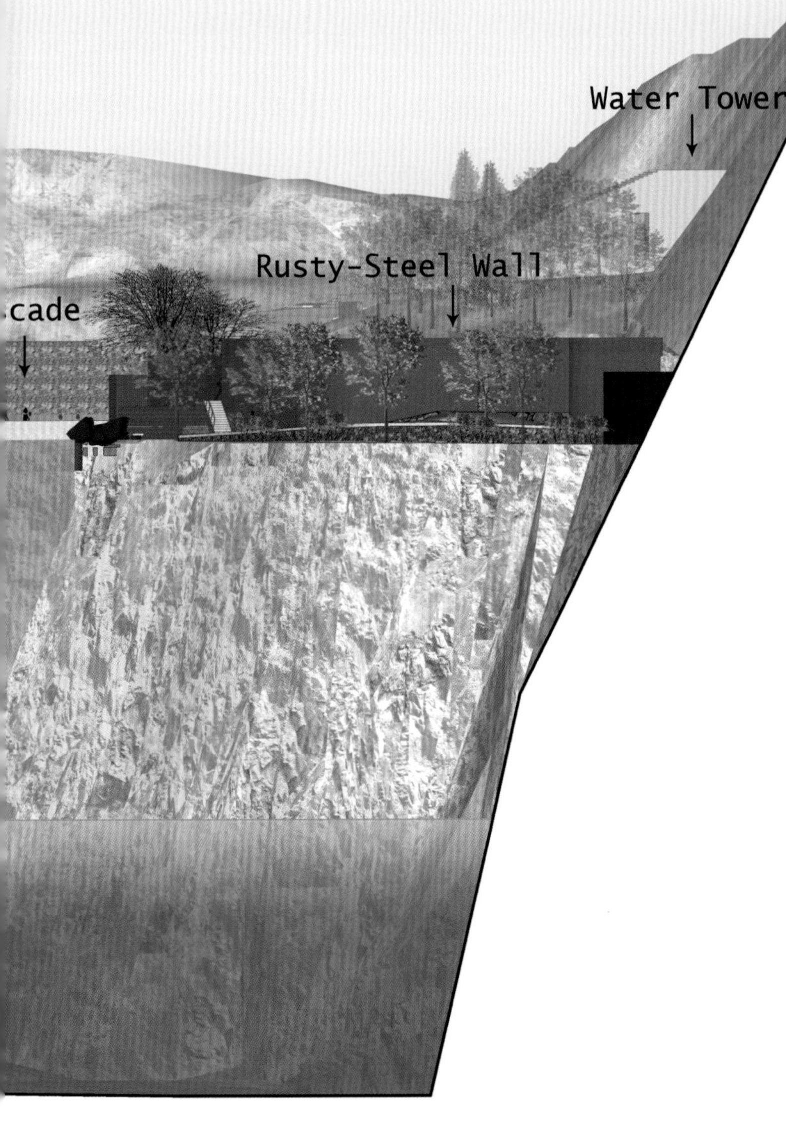

on the ecological degradation of the site, as they can result in surface layer vegetation stripping, acute topographical changes, water and soil loss, landscape damage and habitat fragmentation. The quarry Garden covers an area of about 4.3 hectares (Fig.01-02) and consists of four layers of different heights: hill, plateau, terrace and pool. The surface of the hill area is smooth, heavily weathered, and consistent in shape and texture, with the exception of abrupt rectangular openings on the vertical face. On the plateau, the vegetation is flourishing. This areas contains six entrances leading to the cave depots next to the cliffs. The terrace is the flat section left by quarrying, part of which contains a mature red wood forest. The pool area, covers about 1 hectare and sits 52m below the terrace layer. This area contains clean pool water and lively naturally-formed islands and vegetation (Fig.03).

Plenty of challenges were encountered by the designers during the implementation of this project. One was to reclaim the severely degraded ecological environment. This site has little vegetative cover and a low diversity of species along with severe erosion and loss of water. Another challenge was to excavate and make use of the existing landscape in the quarry. Another imperative was to reestablish proper connections between the quarry and the general population (Fig.04).

The designers adopted the strategy of "Addition and

8

9

10

图 08 人工砌筑的墙与原有毛石墙自然相撞，花草掩映其间。
Fig08 The newly-manmade free stones wall meets the natural free stones at the site, matching with thriving flowers and plants.

图 09 墙体的高低围合形成明确的空间指引
Fig09 The different height and straightness of the landscape walls forms a clear space leading.

图 10 卷曲的坡道将游人带上台地
Fig10 Curled ramp to bring the visitors to the platform

图 11 每一片锈钢板百叶都比前一片多转 0.5 度，形成了梦幻的光影变化，植物从缝隙中钻出来，生机勃勃。
Fig11 Each rusty steel fence inclines 0.5 degree more than the last one, creating dreamlike light change; the plants popped their heads from gaps, luxuriantly and lovely.

图 12 锈钢筒像是要把游客倾倒进湖里，远远地望去，山瀑一泻而下
Fig12 The inclined rusty steel tube takes visitors to the deep pool while hearing the waterfall from far away.

图 13 落日余晖中的锈钢板墙
Fig13 Rusty steel wall In sunset.

图 14 钢筒上切下来的门靠在里侧增强场所的过程感
Fig 14 Process sense that the door cut off from the steel cylinder leans against the inside enhanced place

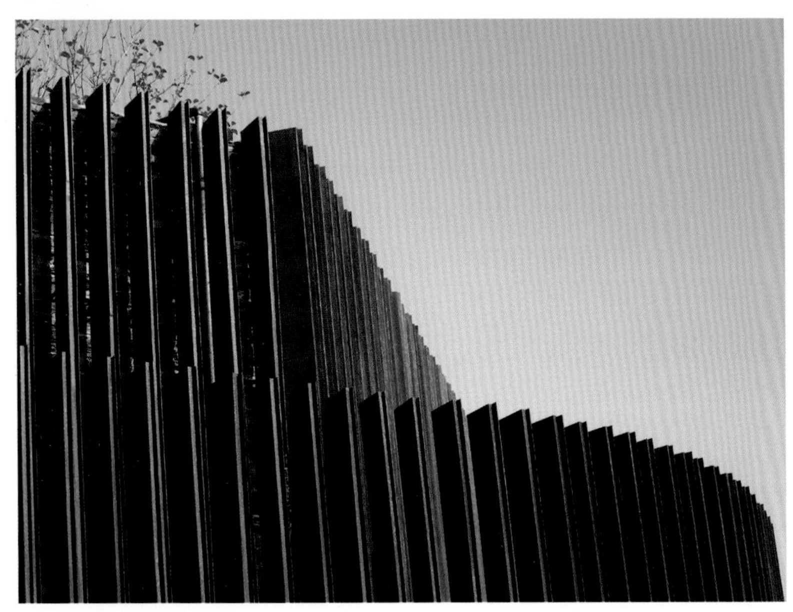

图 15 锈钢板百叶的细部设计
Fig 15 Detailed design of rusty steel louver

图 16 在锈钢板百叶和锈钢板墙交接的地方设置台阶使游人可以上山
Fig 16 Steps are arranged at the joint of rusty steel louver and rusty steel plate wall to enable tourists to climb mounta

图 17 水杉林下的望花台
Fig 17 Flower-seeing platform under metasequoia plantation

图 18 栈道和钢筒看似危险实则非常安全
Fig 18 Trestle and steel cylinder seem as dangerous, but very safe

图 19 蛇形挡墙转弯处种了一株姿态舒展的老梅
Fig 19 Old profile stretch prune tree at the turning of snakelike retaining wall

对应水塔，在镜湖另一侧坡地顶端设置望花台（图 14），可以在镜湖的水光中看一年四季山景变幻。

同时，在东侧山壁之上，开辟出一条山瀑，水从山顶一泻而下，与岩石撞击时带来美妙的水流声。呼应山瀑，援引中国古代“桃花源”（一种典型的隐逸思想，脱离俗世，隐入自然）的意境，顺序设置钢筒（利用悬崖的危险之势，模仿采石时的爆破之态，以倾倒之态势将游人引入栈道）— 栈道（图 15–17）（在行走之际观赏采石留下的山石皴纹，耳畔是山瀑的声响）— 一线天（从采石残留的卷扬机坡道上开辟而来）这条惊险的游线，通过蜿蜒的浮桥（中间的平台可以让游人感受山水交映的美）进入山洞，穿过隧道便来到世外花园。这条游览路线，既精彩刺激又宁静怡人，各种自然之态均含纳其中。

在生态修复与文化重塑的策略基础上，通过极尽可能的链接方式，场地潜力（capabilities）得到了充分表现。一处危险的、不可达的（inaccessible）的废弃地已经转变为使人们亲近自然山水、体验采石工业文化的充满吸引力的游览胜地（图 18–22）。□

Subtraction" as the design principle for the ecological reclaimation for the special forms of the quarry. “Addition” is performed by remolding the topography and increasing the vegetation, thus creating a new biocoenosis. As for the exposed hills and rock walls, the designers managed to respect the trueness of the rock-wall landscape, rather than applying the routine wrapping method. Under the premise of effective avoidance for safety consideration, the designers adopted the "Subtraction" strategy of no intervention, leaving the rock wall to reclaim itself through natural processes. As for the existing retaining wall on the edge of the plateau, the designers used corten steel plates—a material with an industrial character—to wrap it and turn it into a landscape interface featuring the rhythmic variation of sunlight and shade (Fig.05-13).

Enlightened by Chinese landscape painting and classical

图 20 蜿蜒的浮桥上时时可以看到瀑布
Fig 20 Waterfall often seen on winding pontoon bridge

图 21 锈钢板墙和色叶植物相互辉照
Fig 21 Rusty steel plate wall and color-leaf plant happen the same time

图 22 鸟瞰效果图
Fig 22 Rendering Effect of Aerial View

literature, the designers adopted modern design techniques, which reinterpret Eastern landscape culture and Chinese utopianism. The Eastern tradition focuses on the visible and visitable 'accessible' landscape experience, which differs from the Western 'static' landscape tradition. The designers created a "mirrored lake" in the terrace, which can be viewed from all sides, so as to reflect the graceful curves of hills and expand the sense of space. As the vertical face of the hill is somewhat monotonous, a water tower is built against the hill to effectively adjust its rhythm. As well, the addition of flowing spring water from the hill adds a sense of liveliness. Opposite to the water tower, a Flower viewing platform has been set on the top of the slope on the far side of the mirrored lake (Fig.14), where the scenery can be viewed as it changes year round.

A waterfall has been opened on the east hill wall, pouring from the top and crashing against the rocks, with the wonderful sound of falling water. Responding to the waterfall and based on the artistic conception of the "Peach Blossom Garden" (relating to the traditional concept of the hermit living in remote nature), an adventurous walking route has been arranged which highlights the danger of the cliffs and presents an area which recreates the rubble of explosions, and leads tourists into a trestle employing a sequence of steel cylinders (Fig.15-17). The hillstone ripples left from quarrying are echoed by the sound of the waterfall, which can be heard while walking across the "Narrow Sky", a walkway constructed from a winch ramp left from quarrying.

Tourists can come to the Peach Blossom Garden after crossing a winding pontoon bridge, from which the intermediate terrace allows tourists enjoy the beauty of the reflected hills on water, a cave and finally a tunnel. This tourist route is not only spectacular, exciting and pleasant, but also depicts the multifaceted industrial and natural history of the site.

Based on the strategy of ecological reclamation and culture reconstruction, the many different aspects of this site and its history are accessible by making use of pre-existing modes of spatial links. Thus a formerly dangerous, inaccessible, and discarded piece of land has been rebuilt into an attractive recreation area for visitors to approach the natural landscape and experience the culture of the quarrying industry (Fig.18-22).■

作者简介：

翟薇薇\女\景观设计师\北京清华同衡规划设计研究院\中国北京

Biography:

Weiwei Zhai \ Female \ Landscape Architect \ Beijing Tsinghua Tongheng Planning& Design Institute L.td \ Beijing,China

查尔斯·沙（校订）

English reviewed by Charles Sands

丹麦欧登塞斯缇岛垃圾填埋场的景观再现

TRANSFORMATION OF LANDFILL TO RECREATIONAL LANDSCAPE STIGE ISLAND,ODENSE,DENMARK

普莱本·斯卡沃普　泰勒·麦德森　Preben Skaarup　Trine Lybech Madsen

位　　置：丹麦，欧登塞，斯缇岛
面　　积：510000m^2
委 托 人：欧登塞市政局
设计团队：普莱本・斯卡沃普，迈克・约根森，
摩根斯・杜尔摩和安娜・斯卡沃普
建设时期：2007–
完成年代：在持续发展中
获　　奖：在设计竞赛中获一等奖

Site: Stige Island, Odense, Denmark
Area: 510000 m^2
Client: The Municipality of Odense
Design team: Preben Skaarup, Michael Hammelsvang Jørgensen, Mogens Dueholm and Anne Vium Skaarup
Constructing period: 2007 -
Completion year: Under constant development
Award: First Prize in an invited competition

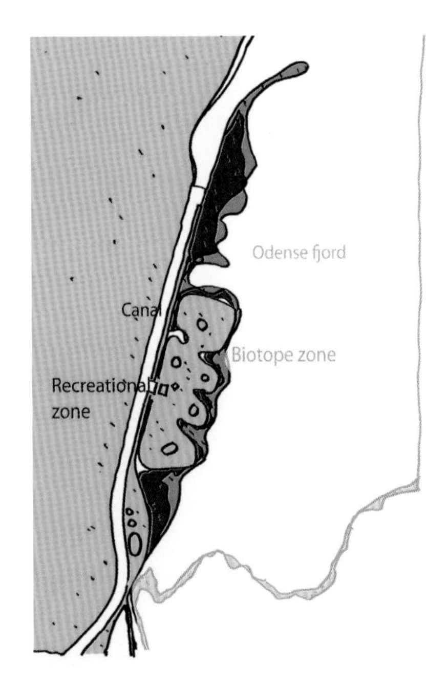

图 1a 分区
Fig1a Zone

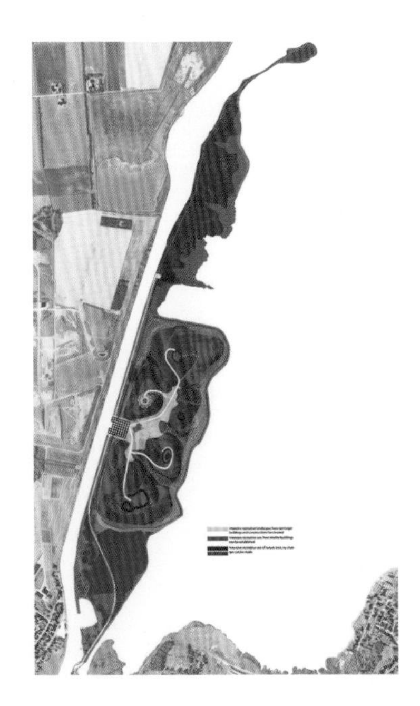

图 1b 策略
Fig1b Strategy plan

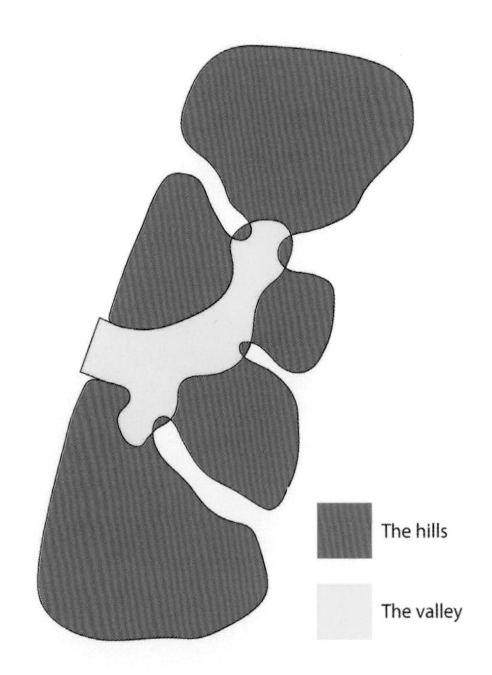

图 1c 山和谷的关系
Fig1c The hills and the valley

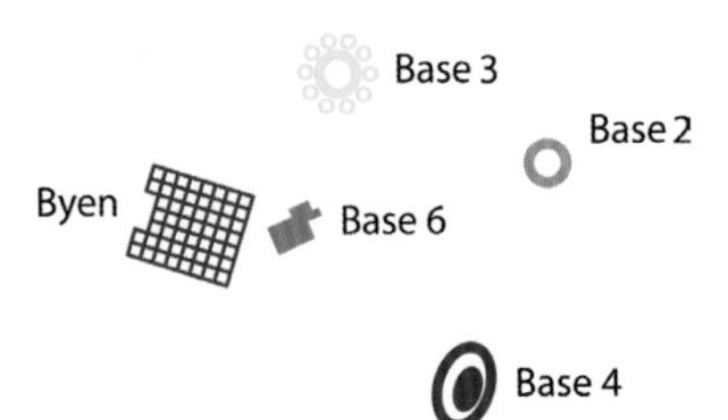

图 1d 基地
Fig1d The bases

图 1e 自然和群落生境图示
Fig1e Diagram nature and biotopes

Asphalt road
Gravel road
Nature path
Path

图 1f 基础设施图示
Fig1f Diagram infrastructure

图 01 概念图
Fig 01 Concept diagram

注：文中所有图片由普莱本·斯卡沃普景观事务所提供
Photography: All images sources from Preben Skaarup Landskab

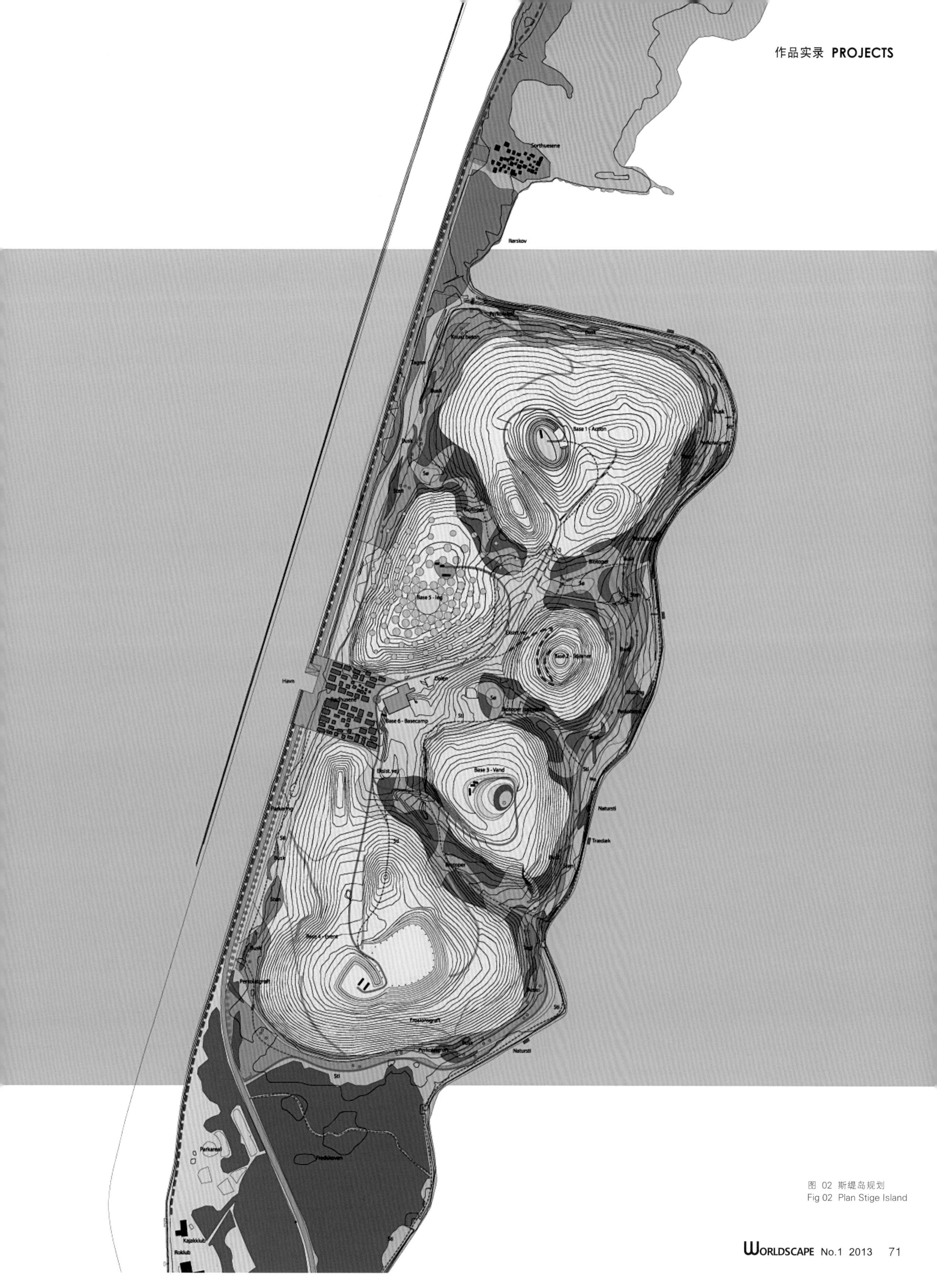

图 02 斯缇岛规划
Fig 02 Plan Stige Island

图 03 活动基地
Fig 03 Collage of the action base

位置介绍

斯缇岛位于丹麦的第三大城市欧登塞以北。岛屿横亘于欧登塞峡湾和欧登塞运河之间。在斯缇岛上，垃圾填埋场的历史可以追溯到1803年，那时欧登塞运河已经建设。岛屿由挖掘运河的剩余土地建成，并且位于一片狭长的土地上。当基址1967~1994年作为垃圾填埋场使用时，岛屿发展起来。垃圾填埋场于1994年关闭。

在以后的日子里，斯缇岛被用作多种用途，例如休闲目的地、船舶制造点、营地等。

2005年的景观竞赛成为转变的节点，垃圾填埋场转化为再创造的景观。普莱本·斯卡沃普景观事务所赢得了竞赛，斯缇岛的发展从此开始了。

斯缇岛的战略规划

斯缇岛的景观是一个令你体验自然、体验地平线和感受巅峰的地方。主要地势由5座高耸的山峰和一组较低的小山组成。在山之间，

Introduction to the site

Stige Island is locatede north of Odense, which is the third biggest city in Denmark. The Island lies between Odense fjord and Odense Canal. The history of the landfill at Stige Island goes back to 1803, when Odense canal was constructed. The Island was made of the surplus land, from the construction of the canal, and was laid out in a narrow strip of land. The Island grew when the site from 1967-1994 was used as a landfill. The landfill was closed in 1994.

Over the years the Stige Island has had multiple functions such as Tivoli site, leisure destination, spot for boat building business, camping site etc.

In 2005 an architectural competition was the starting point of the transformation, from landfill to recreational landscape.

图 04 事件基地
Fig 04 Collage the event base

图 05 红房子基地
Fig 05 Collage of the red houses

宽阔的山谷连接着山、运河和峡湾。

这里的景观是年轻的、不成熟的、不稳定的、未结束的，并且有一整套开放的可能性。由于所有的可能性，挑战出现了：对于新的选择如何保持已有空间的活力和保持现有景观的开放性？如何使景观保持一种正在进行的发展状态？这对于斯缇岛来说，是战略的一个重要的部分，以至于斯缇岛最终可以保留景观发展的机遇。

斯缇岛的战略性规划明确了空间组织、总体计划和基本设施。规划必须简明而精确，同时规划必须适应性强以满足永无止境的变化，也要满足随着时间的推移而改变的要求。

规划必须满足娱乐和自然的双重需要，必须在这两者之间寻求一种平衡，使它们两者互不干扰，但又为游人增加不同的感受。

为了建构这两个主题，岛屿被分成两个区域——一个群落生境区域和一个娱乐区域。这两个区域使斯缇岛与欧登塞城市联结，并且有联结城市、景观和峡湾的功能（图 01）。

所有的现存的和新的娱乐活动，被安排在娱乐区欧登塞运河以西的地带，并且所有潜在的和新的自然体验被安排在欧登塞峡湾的东部的一个群落生境区域（图 02）。

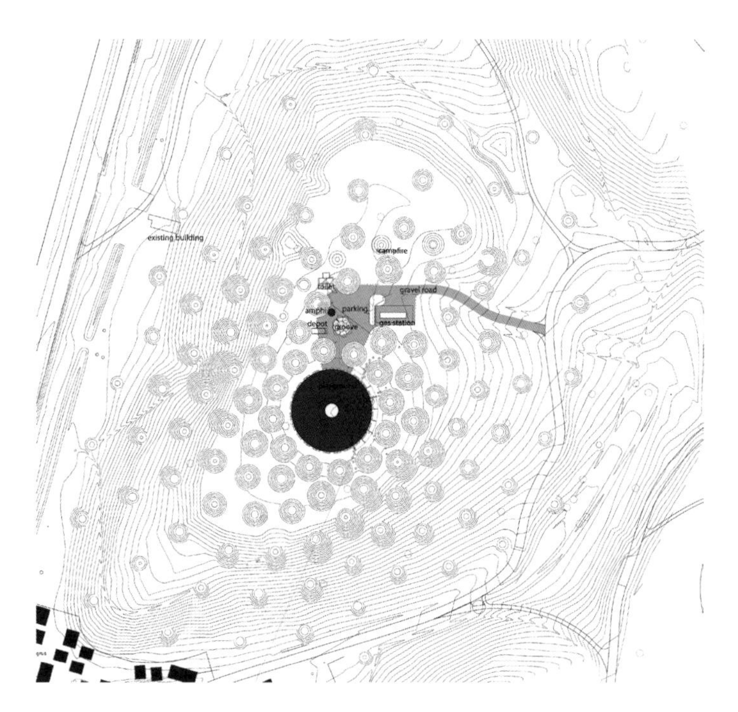

图 06 游戏基地规划
Fig 06 Plan play base

Preben Skaarup Landscape won the competition and the development of Stige Island started.

Strategic plan for Stige Island

The landscape at Stige Island is a place where you can experience the nature, the horizon and the feeling of being on top. The main topography consists of five dominant hills and a set of smaller hills. In between the hills a broad valley connects the hills, the canal and the fjord.

The landscape is young, immature, unstable, unfinished and opens up for a whole set of potentials. Because of all the possibilities the challenge emerged, how to unfold the possibilities, make spaces for already defined activities and still keep the landscape open for new options. How to keep the landscape in an on-going development? This has been an important part of the strategy for Stige Island, so that Stige Island in the end can remain a landscape of opportunities.

The strategic plan for Stige Island defines the organisation of space, the overall program and the basic infrastructure. The plan has to be simple and precise, at the same time the plan has to be flexible and open to the never-ending transformation and adaptions to different demands over time.

The plan has to meet the demands for both recreation and nature. There has to be a balance between the two themes, so they don't disturb each other, but complement each other and enhances the different experiences in a positive synergy.

To structure the two themes, the island is divided in to two zones, a biotope zone and a recreational zone. The two zones connects Stige Island to the city of Odense and functions as a link between city, landscape and fjord (Fig 01).

All existing and new recreational activities are connected in the recreational zone along Odense Canal to the west and all possibilities for existing and new nature experiences are connected to the east in a biotope zone along Odense fjord (Fig 02).

Recreational zone

In the recreational zone bases for activity and play are laid out in the landscape. The bases are situated on top of the 5 large flat-topped hills and the landscape around the bases is modulated so the bases are more or less invisible in the landscape. The defined bases are; the action base, the starwatching base, the freestyle base, the event base and the play base.

图 07 游戏场地
Fig07 The playbase

图 08 游戏基地的棍状物
Fig 08 Sticks at the play base

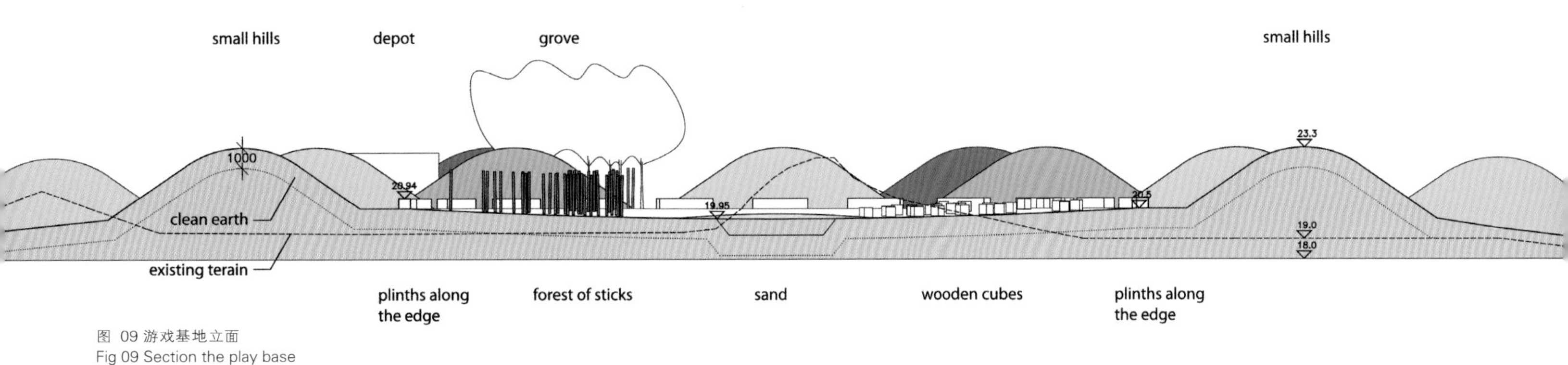

图 09 游戏基地立面
Fig 09 Section the play base

娱乐区

在娱乐区，活动和游戏的基地被设计在景观之中。这些基地被安排在 5 座山的山顶，并且调整这些景观环绕下的基地，使它们或多或少地隐蔽在景观之中。已定范围的基地是这样，活动基地、观星基地、自由式基地、事件基地和游戏基地都是如此。

在山谷中，依着运河，安排了红房子基地。

不是所有的基地都在今天开发，当寻找到新的项目有新活动需求时，未来新的基地将被建立。今天，已经开发了游戏和自由式基地被发展的是那些靠近红房子的基地（图 03–04）。

红房子

红房子是一系列供区域内的人使用的建筑。由小红房子组成的村落像可以获得不同功能的小的集装箱。大广场承担着红房子的集会地点功能，从这里你可以步入斯缇岛的景观之中（图 05）。

In the valley, along the canal, lies the base of the red houses.(Fig 03-04).

Not all of the bases are developed today, but in the future new ones will be established, when the request for new activities and the finances for new projects are found. Today the red houses, the play base and the freestyle base are developed. The developed bases are the ones that lie closes to the red houses.

The red houses

The red houses are a cluster of buildings that are used for supporting the users of the area, the village of little red houses is like small containers that can obtain different functions. The large square connected to the red houses functions of meeting point, from here you go out in to the landscape of Stige Island(Fig 05).

图 10 儿童在游戏基地中
Fig 10 Kindergarten visiting the playbase

图 11 群落生境区域
Fig 11 Collage of the biotope zone

游戏基地

游戏基地是一个环形场所，四周由小山环绕。在大的、微小的直径 40m 的钵形用地上，有一个大立方体、水、沙子、棒状物吸引着来访者游戏（图 06–08）。小山环绕着的基地被用作在景观中的游戏基地的一个自然游戏元素、眺望点和制造者。山有 3m 高，靠近钵形用地，并且比开放的景观低（图 09）。游戏基地是一个自然的参观点，这个参观点是为那些喜欢游戏的儿童和成人准备的（图 10）。

自由式基地

最后一个已经完成的基地是自由式基地。在这块基地上，两个艺术家开发了一种可以酷跑和体育运动的景观。当你登上山，到达基地的所在地，顶部的小混凝土块出现了，当你爬山时，混凝土块可以用来从一点跳到另一点。在平顶山的顶上，一种弯曲的金属管景观出现了，这些管子可以用来跑酷运动和挑战身体。

群落生境区域

斯缇岛的景观，已经被一系列不同的植物居住，但是在斯缇岛开始创建一种强有力的变化万千和多样的群落生境的地域将被结合起来。通过增加一个小数量的新植物物种的群落生境，新的不稳定将出现。通过部分重叠不稳定性，一幅不同群落生境的马赛克画面将成为最终结果。原理很简单，但结果将是不同的和复杂的。通过增加一点，一个新景观将来到生活中（图 11）。

创造结合地带似乎更像自然，结合将作用于马赛克图案，这些图案参考了伪装图案。这些图案具有美感，但仍然具有像自然一样的感觉。马赛克图案的结果是一个景观的演出最终将成为斯缇岛景观的自然部分（图 12）。

与新植物马赛克图案在一起，小的介入发生在景观中。这些介入可以是对于人们、植物和动物的。对人们的介入是景观中公园的长凳和小而低矮的座椅区域。对动物和植物而言，是新的居住者被引入了。

The play base

The play base is a round spot surrounded by small hills. On the large, slightly bowl shaped circle, with a diameter of 40 metres, there are big cubes, water, sand and sticks that invites the visitors to play (Fig 06-08). The small hills around the base are used as a natural play element, lookout points and as a marker in the landscape for the play base. The hills are 3 meters high nearest to the bowl and are lower as they move out in to the open landscape (Fig 09). The play base is a natural visiting point for children and adults that like to play(Fig 10) .

The freestyle base

The last base, that is finished, is the freestyle base. On this base two artists has developed a landscape for play, parkour and physical activity. As you go up the hill, where the base is located, small tops of concrete emerges. The concrete tops can be used for jumping from one spot to another while you are climbing the hill. On the top of the flat-topped hill a landscape of bended metal pipes appears. The pipes can be used for parkour and physical challenges.

Biotope zone

The landscape on Stige Island was already inhabited with a range of different plants, but to start the development of creating a stronger biotope of biological diversity and variation at Stige Island the terrain will be grafted. By adding a small number of new variations of plants to the existing, new variables will emerge. By overlapping the variables a mosaic of different biotopes will be the end result. The principle is simple but the result will be diverse and

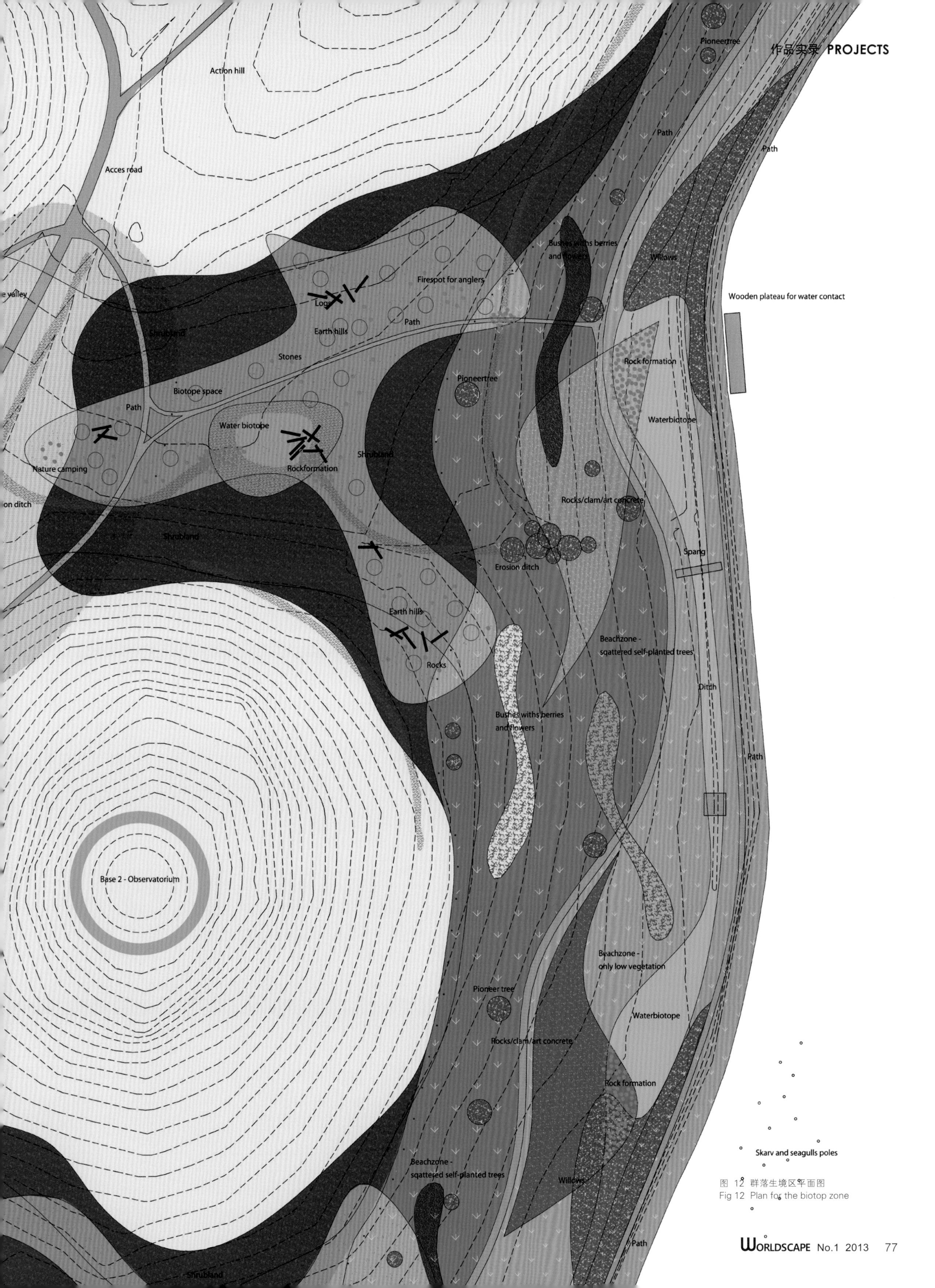

图 12 群落生境区平面图
Fig 12 Plan for the biotop zone

图 13 群落生境区域介入
Fig 13 Intervention – Biotope zone

图 14 群落生境区域的大木头
Fig 14 Intervention – Logs in the Biotope zone

图 15 群落生境区域的成堆的小石头
Fig 15 Intervention – Pile of small rocks in the biotope zone

居住者吸引了小动物和植物的介入位置，它们或者是成堆的较小或较大的石块，或许是有大木头的区域。

在生境群落区域，一个新的碎石路径被引入。该路径联系着植物图案和介入者。所有这些元素创造了一个生境群落区域，这一区域促使人们享受景观中的许多自然过程（图 13–15）。

基本设施

在斯缇岛上，有三层基本设施：道路、小径和自然小路。这些路径铺以沥青或碎石，以满足不同的需求和气氛。路径将为访问者提供步行或骑车环绕斯缇岛的通路，并借此联系城市与景观和基地、上与下、运河与峡湾、山谷和山（图 16）。

新的道路联系着已经存在的道路并通往基地，同时是沼气生产的检查道路。

沿着峡湾的自然路径提供了一种在斯缇岛上沿着靠近山脚下的水体与自然移动的可能。路径由碎石建成，以保留这一区域的自然面貌。

在娱乐区，一种积极的柏油路径被建设在山之间的谷地里，这一路径联系着通往不同基地的小石子路径，并且提供人们骑车、滑旱冰和滑滑板等的区域的通路。在这个路径上的图示指示牌使你有可能衡量你的路线并去做间歇的练习（图 17）。

垃圾填埋场控制

在娱乐性景观的下面，来自垃圾填埋场的剩饭菜仍然是现实的问题。每一英寸的污染物的土地已经被新的土壤覆盖，以阻止人们接触污染物。环绕整个基址，气体管道被埋藏在山下，这样，由于垃圾的腐烂所带来的气体可以被用于发电和生热。在游戏基地的北面，可以看到正在燃烧的由土壤中生成的气体火把。□

complex. By adding a little, a new landscape will come to life(Fig 11).

To make the grafted terrain seem more nature-like, the grafting will be done in patterns, which has references to camouflage patterns. The pattern is aesthetic, but still has a nature-like feel(Fig 12).

The result of the mosaic is a staging of the landscape that over time will become a natural part of the landscape at Stige Island.

Along with the new plant mosaic small interventions are laid out in the landscape. The interventions can be for people, plants and animals. The interventions for people are benches and small lowered seating areas in the landscape. For the animals and plants new habitats are introduced. The habitat attracts smaller animals and plants to the intervention sites; these are either piles of smaller and bigger rocks or areas with big logs (Fig 13-15).

In the biotope zone a new gravel path is introduced. The path connects the plant mosaics and the interventions, together the elements creates a biotope zone that inspires to enjoy the many nature experiences in the landscape.

Infrastructure

There are three layers of infrastructure on Stige Island; roads, paths and nature paths. The paths are established in both asphalt and gravel to meet different needs and atmospheres. The paths will give access for visitors to move around on Stige Island both by foot and bike, and hereby connecting; city to landscape and bases, up and

图 17 活动的路径
Fig17 The activity-path

图 16 斯缇岛全景图
Fig16 Collage panorama of the transformed Stige Island

down, the canal and the fjord, valleys and hills (Fig 16).

The new roads connects to the existing roads and give access to the bases, the roads also functions as inspections roads for the gas production.

The nature path along the fjord provides the possibility to move along Stige Island with a close contact to the water and the nature at the foot of the hills. This path is established in gravel to keep the natural aspect in this zone.

In the recreational zone an activity path in asphalt is established in the valley between the hills. The path connects the smaller gravel paths to the different bases and gives access to the area for people on bicycles, rollerblades, skateboards etc. On the path the graphic gives you the possibility to measure your route and to do interval training. (Fig 17)

Containment of the landfill

Underneath the recreational landscape the leftovers from the landfill is still present. Every inch of contaminated land has been covered in new soil to prevent people from getting in contact with the contamination. Around the whole site gas pipes are drilled down in the hills so that the gasses, from the rubbish decay, can be used for electricity and heat. North of the play base you can see a torch burning with help from the gasses in the soil. ■

作者简介：

普莱本・斯卡沃普 / 男 / 风景园林师 / 普莱本・斯卡沃普景观事务所总裁 / 丹麦奥尔胡斯

泰勒・麦德森 / 女 / 建筑师和风景园林师 / 普莱本・斯卡沃普景观事务所员工 / 丹麦奥尔胡斯

Biography:

Preben Skaarup/Male/Landscape Architect/Owner of Preben Skaarup Landscape /Arhus,Denmark

Trine Lybech Madsen/Female/City and landscape Architect/ Employee at Preben Skaarup Landscape/Arhus,Denmark

陈鹭（中译），朱玲（校译）
Chinese translated by Lu Chen, Chinese reviewed by Ling Zhu

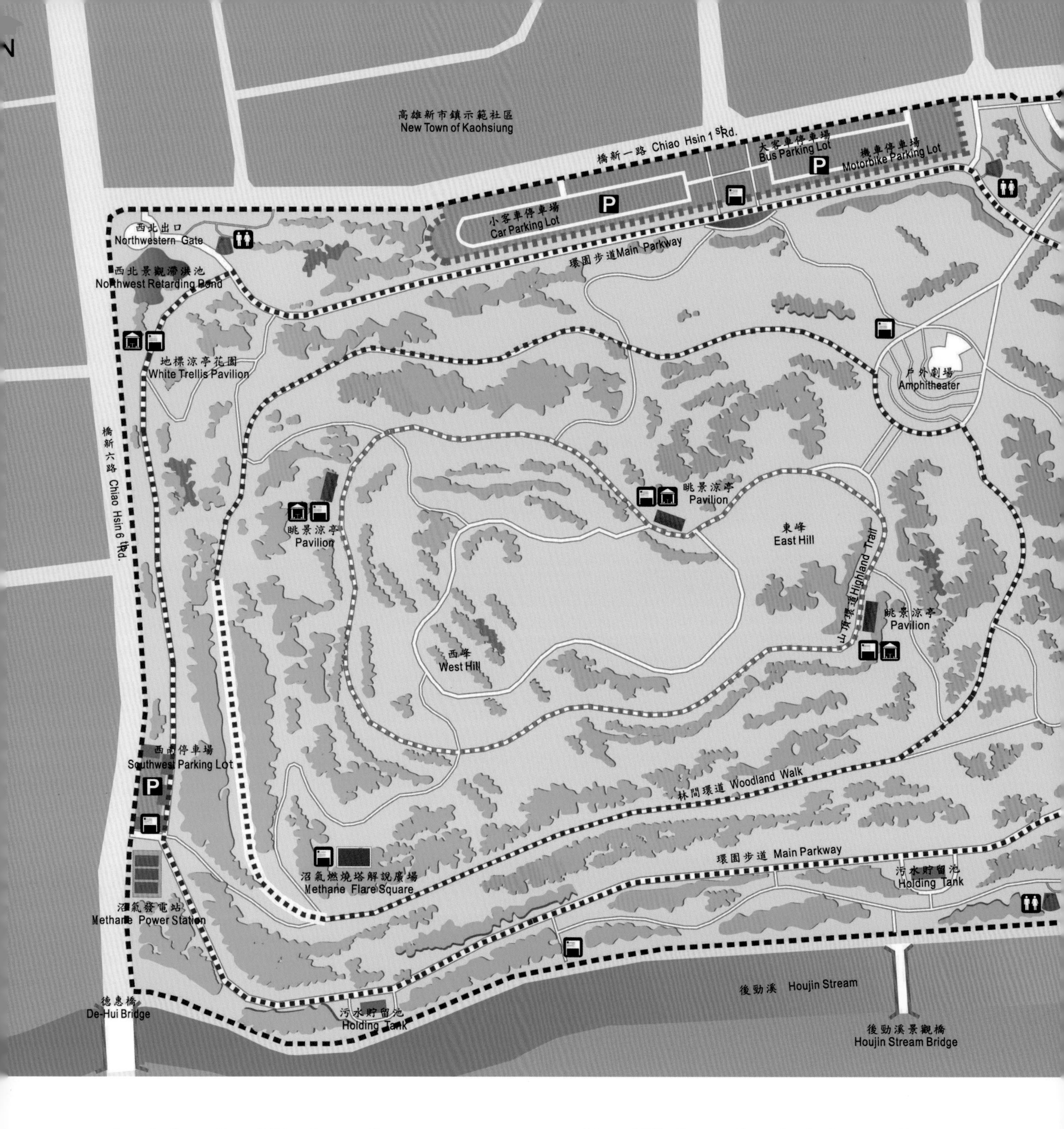

台湾高雄都会公园－发现都会新绿地

DISCOVER A NEW GREEN URBAN SPACE: KAOHSIUNG METROPOLITAN PARK, TAIWAN

汪荷清　洪钦勋　　Heqing Wang　Qinxun Hong

图 01 公园配置图（照片来源：高雄都会公园管理站）
Fig 01 Park Map (Source: Management Section of KMP)

项目位置：台湾高雄市
项目面积：95hm^2
委托单位：内政部营建署
设计单位：皓宇工程顾问股份有限公司 / 行远国际工程开发股份有限公司
设 计 师：汪荷清、邓浩、刘金花、钟艳姒、余添全、黄贵源
完成时间：2009 年

Location: Kaohsiung, Taiwan
Area: 95 hm^2
Client: Construction and Planning Agency, Ministry of the Interior
Designer: Cosmos Inc. Planning & Design Consultants, Progressive Environmental Inc.
Landscape Design: Heqing Wang, Hao Deng, Jinhua Liu, YanShu Zhong, Tianquan Yu, Guiyuan Huang
Completion: 2009

图 02 园区鸟瞰(照片来源：高雄都会公园管理站)
Fig 02 Overview of the Park (Source: Management Section of KMP)

公园的诞生与效益

高雄都会公园位于台湾高雄市楠梓区北端，范围涵跨原高雄县市，面积 95hm²，为高雄都会区居民游憩休闲之最佳场所，也是全台最大的都会公园。园区设计结合都市森林与生态植栽之理念，为国内首座落实环保与游憩的公园，也是首度采用园内设置苗圃生态培育原生种苗木，长成后定植的概念实作计划的基地。全园分为二阶段规划建设完成，第一、二期 35hm² 及 60hm²，分别于 1996 年及 2009 年正式开园启用(图 01–02)。

本园区原为西青浦填埋场，累计填埋垃圾量达 900 万公吨，于 1999 年封闭后，配合复育计划执行，分区分阶段完成填埋区沼气收集设施，以加盖不透水布再行覆土方式，大幅阻绝沼气自表层逸散，提升沼气收集处理成果。每年可妥善处理沼气约 1,900 万 m³，相当于减少 2 万辆汽车行驶 10,000km 的 CO_2 排放量或相当于减少砍伐森林 1.5 万 hm²(图 03–04)。

沼气污染防治厂商以自费方式运用垃圾层中的排气系统收集沼气并设置 5.448MW 沼气发电设施，在营运 20 年间将电力售予台湾电力公司。西青埔沼气处理发电厂一年的沼气回收发电效益可以减少 386,700 公吨的燃煤电厂废气排放量，也使得公园兼具资源再利用、再生能源教育之示范功能。

而山形双峰填埋场最高高度约海拔 47m，使得二期公园用地成为平原中一座地标大山丘，山区俯瞰都会景观绝佳。由于绿化成果良好，林木成荫，设施完善，蔚为大高雄地区环境教育及举办活动之良好场所，2012 年全年约服务逾 120 万人次来园游客，并办理大型活动 20 场次。

公园的景观复育与蜕变

The historical development of the Kaohsiung Metropolitan Park and its benefits

Spanning the boundary between Kaohsiung City and Pre-Kaohsiung County, the Kaohsiung Metropolitan Park is located at the north end of Nanzih District, covering an area of 95ha. It is the biggest metropolitan park in Taiwan and provides the residents of the Kaohsiung Metropolitan area a place for outdoor recreation. It is the first time that a park in Taiwan has introduced the technique of onsite nursery breeding of native species for the purpose of colonization. The park's design incorporates ideals of urban woodlands and ecological planting, making it the first park in Taiwan to fully combine environmental protection and recreation. The park was planned and constructed in two stages, with the first area covering 35ha completed in 1996 and the second area covering 60ha completed in 2009. (Fig01-02).

Previous to the construction of the park, the site contained the Chingpu Landfill, which collected a total of nine million tons of garbage before closing in 1999. As part of the restoration plan the ground was covered with impermeable canvas to prevent methane from being released through the surface and to improve methane collection results. 19 million cubic meters of methane are collected and processed at the site every year, reducing CO_2 by an amount equal to the emissions of 20,000 automobiles or the cutting down of 15 thousand hectares of

图 03 沼气发电厂（照片来源：高雄都会公园管理站）
Fig 03 Methane Power System (Source: Management Section of KMP)

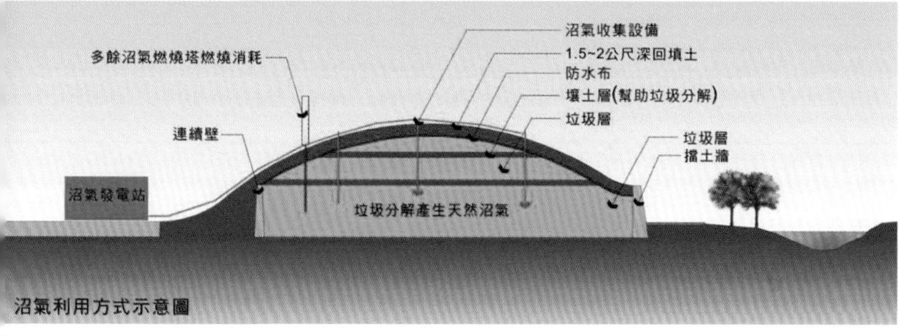

图 04 沼气利用方式示意图（照片来源：皓宇工程顾问有限公司）
Fig 04 Biogas utilization diagram (Source: Cosmos Inc.)

forest (Fig 03-04).

The garbage layer in the exhaust system is used by biogas pollution control manufacturers to collect biogas. A 5.448MW biogas generation facility will operate for 20 years with the electricity generated by the gas sold to the Taiwan Power Company. The same amount of power created from the yearly recovery of biogas from the West Qingpu gas treatment plant would create 386,700 metric tons of emmisions at a coal-fired power plant. Thus the park performs the multiple functions of leisure and recreation, as well as the practical and educational demonstration of resource recycling and renewable energy.

With a high point of about 47 meters, the 2nd stage area of the landfill site acts as a landmark hill on the plains, providing an excellent viewing point for the surrounding landscape. Due to good facilities and maintenance, the park is now the primary destination for environmental education and outdoor activities in the Kaohsiung region. In 2012 the park attracted over 1,200,000 visitors and held 20 mass activities.

Landscape restoration and change

The park is separated into 3 areas: the Entrance Area, the Sport Activity Area, and the Forest and Botanical Area. The timeline for the planning and construction of the two stages of the park is indicated in Table 1. The 1st stage encompassed the planning of the landfill and the Forest park, while the 2nd stage encompassed the restoration of the landfill site and the Forest Botanical Park. In the centre of Forest Park, is a raised terrain of hills, which is the site

图 05 从掩埋场到公园的景观变迁（照片来源：高雄都会公园管理站）
Fig 05 From landfill (1997) to landscape to park (2008) (Source: Management Section of KMP)

图 06 2005 森林园区工程施工现场空照图（照片来源：皓宇工程顾问有限公司）
Fig 06 2005 Forest Park construction site, aerial photographs (Source: Cosmos Inc.)

图 07 第一期园区开放初期（照片来源：高雄都会公园管理站）
Fig 07 Period after the opening of the first stage of the park (Source: Management Section of KMP)

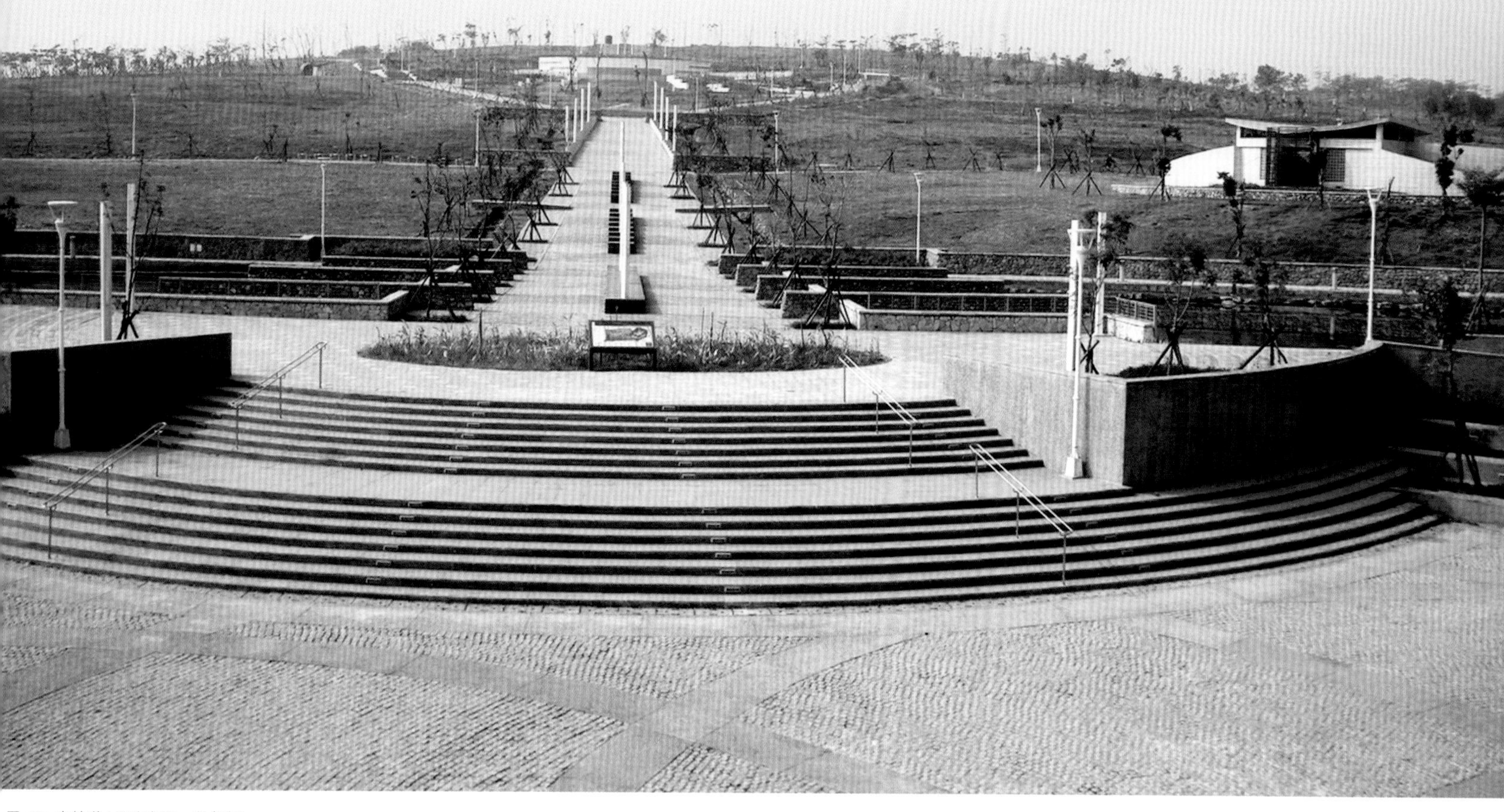

图 08 中轴道（照片来源：高介志）
Fig 08 Main Parkway (Source: Jiezhi Gao)

本园园区分为入口区、动态活动区及森林植物区，其规划建设分为两个时期（表 1）。第一期为垃圾填埋使用及动态活动区规划建设时期，基地平坦，以动态娱乐活动及平面森林公园为主，第二期为垃圾填埋场封场及森林植物区规划建设时期，其森林公园中央为地形隆起的山丘静态休闲散步区，提供眺景回廊、步道、户外剧场等景观设施。全区环状动线由高而低分为山顶环道、林间环道与环园步道；环道东北侧可通往区内非填埋场的活动腹地，包括落瀑广场、荷花池、植物主题花园、儿童游戏场、庆典广场等景观设施区。停车场区规划于北侧及西侧供游客使用，并便于未来北侧新市镇计划区市民利用。环道东侧设置一座人行吊桥、一座人车共享平面桥梁，以联络一期园区管

of walking trials, scenic corridors, and an arena. The peak ring road, the forest ring road and central park trails form a series of circular lines, which move from the high areas to the low areas of the site. To the northeast, the ring road connects to the areas of non-landfill activities in the hinterland. Here are lotus ponds, plant themed gardens, a children's playground, waterfall square, and celebration square. The parking area for visitors is located at the north and west sides of this area to facilitate public access to the site from the newly developed area of Kaohsiung to the north. To the east of the ring road, a pedestrian suspension bridge, and a

表 1 高雄都会公园发展史

阶段	时间	纪事
台糖低洼农地	1971 年代以前	台糖甘蔗田
第一期园区垃圾填埋时期	1977–1984 年	西青埔洼地：倾弃式填埋（旷野倾废处理） 西青埔垃圾场：三明治式填埋
	1985–1990 年	高雄市卫生填埋场
第一期园区动态活动区建设期	1989 年	行政院核定「高雄都会公园开发计划」
	1991/12/15	高雄都会公园第一期园区开始施工与植栽绿化
	1996/04/17	高雄都会公园第一期园区开放使用
第二期园区森林植物区建设概况	1999/06/29	垃圾填埋场封场
	2000/05/29	西青埔沼气处理发电正式启用
	1999–2004 年	持续进行全区填埋
	2005/03	高雄市环保局正式将基地交给内政部营建署继续兴建高雄都会公园第二期园区
	2009/04/19	第二期园区启用

Table 1 The history of the development of Kaohsiung Metropolitan Park

Stage	Time	History
TSC Low–Lying Farm	Before 1971	TSC Sugar Cane Farm
The 1st Stage of Park Period of Use for Refuse Burial	1977–1984	West Chingpu Low–Lying land : Open field dumping West Chingpu Wasteyard : Sandwich landfill
	1985–1990	Kaohsiung Sanitary Landfill Site
The 1st Stage of Park Sport Activity Area – Construction Period	1989	Plans to develop Kaohsiung Metropolitan Park approved by Executive Yuan
	Dec. 15, 1991	Work and planting begins in the 1st stage of Kaohsiung Metropolitan Park
	Apr. 17, 1996	1st stage of Kaohsiung Metropolitan Park opens for use
The 2nd Stage of park Forest Work Area – Work Summary	June 29, 1999	Refuse Landfill Site closed
	May 29, 2000	Methane power facilities for the West Chingpu Marsh Gap goes into use
	1999–2004	Burial of entire area carried out
	March, 2005	The Kaohsiung Environmental Protection Bureau transfers the park to the Ministry of the Interior' s Construction and Planning Agency which continues to construct the 2nd stage of park
	Apr. 19, 2009	The 2nd Stage of park goes into use

图 09 黄连木内环步道（照片来源：刘川）
Fig 09 Park Trails (Source: Chuan Liu)

图 10 落瀑水墙的造型以抽象山水作为主题（照片来源：皓宇工程顾问有限公司）
Fig10 Baoshui waterfall (Source: Cosmos Inc.)

vehicular/pedestrian bridge connect to the management center and the activity area (Fig05-10).

The park is located just to the south of the Tropic of Cancer, a tropical monsoon zone, with an annual mean temperature of 25℃ and an average annual rainfall of 1,700 mm, mostly concentrated in the summer. The park has to face seven months of drought from October to April of the following year. For this climate, drought and heat tolerant vegetation are most appropriate. The park was designed to provide diverse ecological habitats, including different degrees of gentle slopes, peaks and plains, combining areas of gravel, sand, swales, ponds, wetlands, and dry ditches, as well as areas of long grass, short grass, woods and bushes. Careful design techniques were applied for the tropical garden area, providing exhibition and educational opportunities.

Major Facilities Design Concepts

Waterfall square

Under the landscape bridge towards the new town is waterfall square, a semi-circular sunken plaza which acts as an entrance plaza to the park and links the park and Triangle Garden of the new town. To the east of the sunken plaza, a waterfall is created, which makes use of the different levels of the plaza. Natural stone surface treatment and the silhouette of mountain form the image of an abstract waterfall (Fig 11).

Lotus pond

Lotus pond is a detention pond designed for disaster prevention purposes and planted with enriched vegetation. To the north of the pond, a half-moon shaped steel structure is designed together with a hydrolic stepped revetment. The south banks of the pond are vegetated and the central island is planted with water willows. On both sides of the central axis road, a demonstration area of aquatic plants is established. Crossing

图 11 日晷广场（照片来源：高雄都会公园管理站）
Fig 11 Sundial Square (Source: Management Section of KMP)

理中心及绿地活动空间（图 05–10）。

由于公园地处北回归线以南，属于热带季风区，年平均气温为摄氏 25 度，年降雨量为 1,700mm，但多集中在夏季，每年由 10 月到翌年 4 月为长达 7 个月的干旱期，较适宜耐高温、耐旱的植物生长。公园在规划阶段即尽量提供多样化生态栖地，因此，公园中的生物栖地类型包括平缓度不同之坡地、山顶草原、平地草原、砾石地、碎石地、沙地、深浅不一的洼地、池塘、湿地、干沟等，并以植栽设计手法营造长草区、短草区、树林、灌木丛、诱蝶诱鸟等食源植物，并于热带花园区种植南部低海拔区及海岸林带常见植物，提供展示及教育机会。

主要设施区设计说明

入口落瀑广场：在与新市镇相接之景观桥下方，以半圆形之下凹式广场设计手法，将高雄都会公园与新市镇的三角公园，做空间上的串连，作为将来由新市镇进入本园区之入口广场。广场东半边利用下凹式广场的高低差设计落瀑水景，以天然石材之质感与山形剪影层次变化，塑造抽象瀑布意象（图 11）。

荷花池：利用公园防灾计划上必要的沉沙滞洪池，美化成安全且植栽丰富的荷花池。配合沉沙需求为钢性结构，形状接近半月形，北侧池岸设计为亲水阶梯护岸、南侧池岸为植生护岸、水中塑造植生小岛种植水柳，中轴道两侧设置水生植物教学区，水面上跨设栈桥连接落瀑广场与花园区（图 12–13）。

湿地花园区：由荷花池出水口水道引水至湿地区，池底为黏土，池岸四周种植各种形态的水生植物，以台湾本地物种为主，可供市民认识植物（图 14）。

儿童游戏场：提供符合不同年龄层次需求之儿童游戏设施，有组

图 12 荷花池种植多种水生植物可净化水质，并且作为环境教育教材（照片来源：皓宇工程顾问有限公司）
Fig12 Lotus pond with a variety of water purifying aquatic plants (Source: Cosmos Inc.)

图 13 千溪流过的地方以跨桥作为环山步道的串连（照片来源：皓宇工程顾问有限公司）
Fig 13 Meandering mountains trails with bridges crossing Qian stream (Source: Cosmos Inc.)

合游具、攀爬设施、溜滑梯、游戏桥等，地坪种类上有安全地垫以及沙坑两种不同的质感。着重立体空间连序配置的趣味性。

庆典花园广场：是全园最吸引眼球的大型休憩设施区，配置于园区东北侧、非垃圾填埋区且较为平坦之腹地。设置椭圆形草坪广场、7 座白色大型遮阳造型花架、花台坐阶等，步道材质采用天然砂岩与黄褐色系板岩。提供市民团体休闲庆典活动的地点。

热带花园：由于园区地处台湾南部热带地区，植物花园主要展示南部低海拔区、海岸林区、季风雨林区、隆起珊瑚礁代表性植物为主题；外围搭配热带地区常见的景观性开花植物，展示花台以减少水泥用量之石笼工法施作（图 15）。

户外剧场：设置于山区东峰山腰处，规模约 25m×25m 的舞台区提供表演及准备空间，看台区则运用坡地地形规划为草坡坐阶形式，全区含外围步道约为直径 90m 之圆形，本区视野良好，可鸟瞰园区花园区景观及新市镇腹地（图 16）。

山顶草原区：本公园山顶整理为平坦草地，提供大面积的草原活动空间，沿山顶步道设置 3 座木作眺景凉亭，兼具休憩与眺景功能（图 17）。

沼气燃烧塔解说广场：针对既有沼气燃烧塔设备进行美化，并设置解说小广场，提供沼气相关信息，达到资源再利用、再生能源教育之示范。

缤纷多样的生态

本园区林木蓊郁，栽种植物达 10 万株、400 多种，以原生本土植物为主。为了提供多样的生物栖息环境，植栽以生态原则进行栽种设计，采用复层林植栽方式，保留树林的隐密和结构，多层次的栖位空间，提供各类生物多样的食源及不同季节开花与果实，丰富的蜜源植物和食草，吸引种类多样的昆虫摄食、繁衍（图 18–19）。

the water, a trestle is set up to connect waterfall square and the garden area (Fig12-13).

Wetland garden

The water for the wetland garden area, which is paved in clay, is provided by an outlet from Lotus pond. Around the pond's shore are various types of aquatic plants, mainly native species, which also serve to educate the public (Fig 14).

Children's playground

The children's play facilities are provided to meet the needs of different age groups. A variety of playground equipment such as climbing facilities, slides, and game bridges are combined with safety mats and sand pits providing different surface textures and delivering an interesting three-dimension space.

Celebration square garden

Located on the northeastern part of the landfill area and a relatively flat area in the hinterland, Celebration square is the largest and most attractive recreational facility in the park. Seven large white sunshading flower pergolas and flower beds are placed on an oval lawn. Natural sandstone and slate are used for the trails, providing a place for leisure and celebratory events for the public.

Tropical garden

The tropical garden displays the representative plants of southern low-altitude regional coastal forests, monsoon rainforests, and exposed coral reefs. Gabion construction methods were applied. On the edge of the garden, the flowering plants of

图 14 将多余的雨水收集到开放式或地下式滞洪池（照片来源：皓宇工程顾问有限公司）
Fig 14 Excess rainwater is collected in open water features and in underground detention ponds (Source: Cosmos Inc.)

图 15 热带花园区以多孔隙的石笼做花台兼坐椅（照片来源：皓宇工程顾问有限公司）
Fig 15 The Tropical Garden area with gabion planters, flower beds and benches (Source: Cosmos Inc.)

图 16 由落瀑广场环视山坡区，沿中轴廊道跨过荷花池，可接到顺应坡地地形设计的露天音乐剧场（照片来源：皓宇工程顾问有限公司）
Fig 16 Waterfall Square with lotus pond area, and corridor connecting to the open-air music theater, which conforms to the sloping terrain (Source: Cosmos Inc.)

图 17 园区大草原近况（照片来源：高介志）
Fig 17 The park meadow (Source: Jiezhi Gao)

18

19

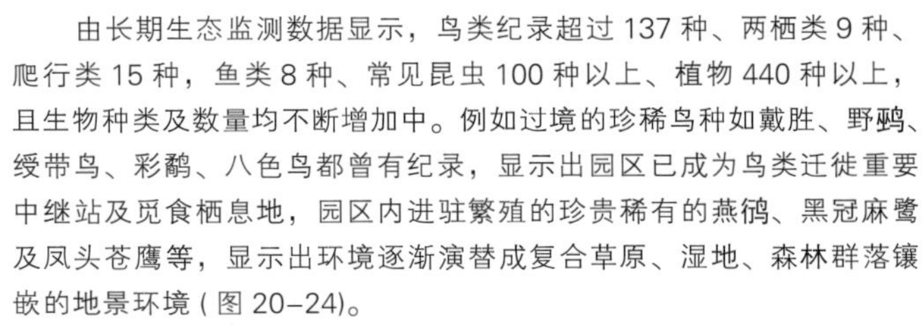

由长期生态监测数据显示，鸟类纪录超过 137 种、两栖类 9 种、爬行类 15 种，鱼类 8 种、常见昆虫 100 种以上、植物 440 种以上，且生物种类及数量均不断增加中。例如过境的珍稀鸟种如戴胜、野鹀、绶带鸟、彩鹬、八色鸟都曾有纪录，显示出园区已成为鸟类迁徙重要中继站及觅食栖息地，园区内进驻繁殖的珍贵稀有的燕鸻、黑冠麻鹭及凤头苍鹰等，显示出环境逐渐演替成复合草原、湿地、森林群落镶嵌的地景环境 (图 20–24)。

未来展望

高雄都会公园提供了垃圾填埋场活化再利用的价值，不但利用原本的垃圾产生沼气来发电，最高纪录可供应 7000 户居民日常所需，节能减碳，也利用垃圾堆起的高低落差地形，创造了人与自然双赢的生态空间，为环境改造利用与都市生活空间增添一股生命力 (图 25)。

园区丰富的生态环境与生物多样性，结合素质优异、专业且人力充沛的志工，提供各级学校户外生态体验与环境设计的重要学习场域，及未来周边高雄新市镇的规划，创造优质城市生态环境，成为高雄都市之肺，提供社会大众体验健康生态之旅 (图 26)。

园区的维护管理采用生态原则方式处理，不喷洒农药、设置枯枝落叶堆置区，保持落叶层的完整性和有机物的循环，提供小型动物栖息环境，并规划森林、灌木、草原、水塘、溪流、荒地等多样生物栖息环境，目前持续进行园区的生态监测，营造生态新栖地乐园。另外，

tropical regions are arranged to constrain the flower beds and reduce the amount of cement used in the construction. (Fig 15).

Arena

The stage area with a size of 25 x 25 meters is set into the mountainside providing a unique place for performances. The grassy area with its sloped terrain is used for seating. The entire area is surrounded by a circular trail, with an approximate diameter of 90 meters, providing a comprehensive lookout on the garden area of the park and the new towns (Fig 16).

Hilltop prairie

The peak is covered by a flat grass area, providing a large grassland for activities. Three wooden pavilions are set along the peak trail for resting and viewing purposes (Fig 17).

Biogas combustion tower and interpretation square

The biogas combustion tower has a monumental sculptural presence, while a small plaza provides information on biogas and demonstrates the process of resource recycling and renewable energy.

An Abundance of Ecological Diversity

Plant life grows thick in the park, with some 100,000 plants of

为推广「环境教育」，园区将每月举办各项讲座、展览及解说活动。同时为提供都会区民众优质休闲游憩场所，园区将持续改善环境质量和相关休闲游憩设施，以及结合在地资源、建立伙伴关系，并配合本公园通过环境教育设施场所认证，开创环境教育新作为，使本园成为国家公园于高雄都会区的橱窗。□

over 400 species. Most of them are native species. In order to provide several kinds of habitats, stratification has been taken as the main principle in the planting scheme. The density and structure of the forest is preserved, providing multiple levels of vegetation and diverse food sources for various organisms as well as plentiful pollen sources and grasses which bloom and bear fruit in different seasons, attracting numerous species of insects to feed and reproduce (Fig 18-19).

Long-term ecological monitoring has idintified over 137 species of birds, 9 species of amphibians, 15 species of reptiles , 8 species of fish, 100 species of common insects and 440 kinds of plants, with their numbers and the number of species continually increasing. Rare migratory birds such as the Hoopoe (Upupa Epops), Yellow Bunting (Emberiza Sulphurata.), Black Paradise Flycatcher (Terpsiphone Atrocaudata.), Painted Snipe (Rostratula Benghalensis (Linnaeus).), Blue-winged Pitta (Pitta Brachyura Temminck & Schlegel) and others have been recorded here, indicating that the park has become an important stopover point for migratory birds to rest and seek food. The habitation and breeding of rare birds such as the Indian Pratincole(Glareola Maldivarum,, Malay Night Heron (Gorsachius Melanolophus.), and Crested Goshawk (Accipiter Trivirgatus) indicate that the area has gradually become a combination of grassland, wetland, and forest (Fig 20-24).

图 18 鲁花树（照片来源：王凤雀）
Fig 18 *Scolopia oldhamii hance* (Source: Fengque Wang)

图 19 穗花棋盘脚（照片来源：王凤雀）
Fig 19 Small-leaved Barringtonia (*Barringtonia racemosa (L.) blume ex DC*) (Source: Fengque Wang)

图 20 彩裳蜻蜓（照片来源：高介志）
Fig 20 Common Picture Wing Dragonfly (*Rhyothemis variegata aria*) (Source: Jiezhi Gao)

图 21 凤斑蛾（照片来源：梁靖微）
Fig 21 Smoky Moth (*Histia flabellicornis ultima*) (Source: Jingwei Liang)

Future Prospects

Kaohsiung Metropolitan Park has created value by reusing a landfill site. It not only uses the methane produced by buried

图 22 黑冠麻鹭－育雏（照片来源：张荣晋）
Fig 22 Malay Night Heron (*Gorsachius melanolophus*) with brood (Source: Rongjin Zhang)

图 23 野鸲（照片来源：张荣晋）
Fig 23 Siberian Rubythroat (*Luscinia calliope*) (Source: Rongjin Zhang)

图 24 贡德氏赤蛙（照片来源：高介志）
Fig 24 Gunther's Frog (*Rana guentheri*) (Source: Jiezhi Gao)

图 25 景观亲水区（照片来源：高介志）
Fig 25 Water Activity Area (Source: Jiezhi Gao)

refuse to generate power which meets the needs of 7000 residents, but also makes use of the rolling terrain created by the buried refuse to create an environment that benefits both man and nature, adding natural vitality to the city (Fig 25).

The Park's rich ecological environment and diverse living creatures, are explained by numerous professional guides, providing an important outdoor learning environment for schools of all levels. In the future, the park will provide a high quality ecological space, acting as an urban lung for Kaohsiung City and the surrounding area, and providing the public with an opportunity to experience a healthy ecological journey (Fig 26).

Management and maintenance of the park is based on ecological principles. Agricultural chemicals are not used, and fallen branches and leaves are allowed to accumulate in some areas, preserving the leaf cover and the organic cycle and providing habitat for small animals. Forest, shrub land, prairie, pond, brook, open land and other habitats are planned to build up an ecological base. In addition, in order to promote environmental education to the public, a lecture, exhibition and interpretation activity will be held every month. In the meantime, to provide a better quality of recreation and habitat, the management agency will continuously improve the facilities and environment of the park. Furthermore, public welfare organizations have been invited to work with the park. And following up on the award for environmental education recently received from the government, a new regional strategy for the park has been set up. It is expected that the Kaohsiung Metropolitan Park will be a window to connect with the National Park.■

图 26 2012 年元旦民众参加单车悠游绿色世界活动（照片来源：高雄都会公园管理站）
Fig26 An organized bicycle ride on Jan 1st, 2012 (Source: Management Section of KMP)

作者简介：

汪荷清／女／硕士／皓宇工程顾问股份有限公司总监
洪钦勋／男／硕士／高雄都会公园管理站主任

Biography:

Heqing Wang/Female/Master/Director of the Cosmos Inc. Planning & Design Consutants
Qinxun Hong/Male/Master/Director of Management Section of Kaohsiung Metropolitan Park

洪钦勋 何欣慈（英译），查尔斯・沙（校订）
English translated by Qinxun Hong, Xinci He, English reviewed by Charles Sands

岭南园林 美丽中国

对话大师

MASTER DIALOGUE

彼得·拉茨　PETER LATZ

工程文凭，景观建筑师和城市规划师
德国慕尼黑技术大学的荣誉教授

Dipl. Ing. Landscape Architect and Urban Planner
Prof. emeritus of excellence TU Munich

图 01 彼得·拉茨说明北杜伊斯堡景观公园 (照片来源 : Michael Latz)
Fig 01 Peter Latz explaining Duisburg Nord (Source: Michael Latz)

图 02　2005 年彼得·拉茨和夫人安娜丽莎在 MOMA(照片来源：Michael Latz)
Fig 02　Anneliese + Peter Latz at the MOMA 2005 (Source: Michael Latz)

图 03 彼得•拉茨公司人员 (2010 年 10 月) (照片来源 : Latz + Partner)
Fig 03 The team October 2010 (Source: Latz + Partner)

图 04 Tilman 解说展览项目"壤的地方和绿洲" (照片来源 : Latz + Partner)
Fig 04 Tilman explaining our exhibition "Bad Places and Oases" Chamber of architects, Munich (Source: Latz + Partner)

图 05 L+P 团队 (2012 年) (照片来源 : Latz + Partner)
Fig 05 L+P Team 2012 (Source: Latz + Partner)

图 06 北杜伊斯堡 – 铁道公园 (照片来源 : Latz + Partner)
Fig 06 Duisburg Nord – Railway Park (Source: Latz + Partner)

关于彼得 · 拉茨

尤杜 · 魏拉赫（博士，德国慕尼黑技术大学教授）在其所著《景观语法 – 彼得 · 拉茨与其伙伴的景观建筑》一书中提到（ISBN：13：978–3764376147，Birkhäuser 出版）："在过去半世纪里，景观专业领域在概念上与范围上已大幅度的拓展，但是很少人真正能对景观建筑界所面对的复杂问题，做出适切的响应，并且诠释符合社会条件下当代环境设计的新表现形式，就如同当年奥姆斯特德在其时代所面临的挑战而展现的睿智与创造性"。彼得 · 拉茨是当今少数能成功的运用娴熟的技巧于文化上，而有所斩获者，他并致力于后工业时代文化价值的重塑。在国际上他是深具影响力的景观建筑师之一，不仅将其专长展现于实务上，也在德国慕尼黑技术大学、美国哈佛大学和宾夕法尼亚大学作育英才及从事研究工作。他的设计从不呆板，并呈现相当多元的风貌。由于他了解每一个案子都会产生错综复杂的影响，因此更注重从理论与科学为基础出发与切入，致力完成精准的作品。

7

About Peter Latz

Udo Weilacher (Prof. Dr., TUM Technical University Munich) writes in his book ***Syntax of Landscape – the Landscape Architecture of Peter Latz and Partners*** (ISBN : 13: 978-3764376147, Publ. Birkhäuser):

"Over the last five decades the word "landscape" has been greatly extended both conceptually and in scope, but few people have fully understood how to respond appropriately to the considerable increase in complex problems that landscape architecture faces, or indeed how to develop – as Olmsted did in his day - new expressive forms of contemporary environmental design that suit the prevailing social conditions. Peter Latz is considered one of the few representatives of our profession who obviously succeeded in making this cultural breakthrough with his skilful transformation and cultural revaluation of post-industrial landscapes. He is now one of the internationally significant landscape architects acclaimed for his expertise both as a professional practitioner and also for his university research and teaching at institutions including the Technical University in Munich and the Universities of Harvard and Pennsylvania. There is no template for his work. His projects are many and various, and all steeped in his commitment to crafted precision and a sound theoretical and scientific basis drawn from his awareness of the complex range of effects likely to be triggered and characterizing the reality of each project as he finds it."

Peter Latz graduated from the TUM Technical University of Munich in 1964. After four years of postgraduate research and studio work in urban planning at the RWTH Aachen, he founded his own practice as independent landscape architect and

图 07–08 北杜伊斯堡 Metallica 广场（照片来源：Michael Latz）
Fig 07–08 Duisburg Nord Piazza Metallica (Source: Michael Latz)

图 09 北杜伊斯堡—煤仓花园（照片来源：Michael Latz）
Fig 09 Duisburg Nord — Bunker garden (Source: Michael Latz)

图 10 更新后的北杜伊斯堡污水处理厂 (照片来源 : Michael Latz)
Fig10 Duisburg Nord – transformed sewer (Source: Michael Latz)

图 11 北杜伊斯堡 – 公园的秋天景色 (照片来源 : Michael Latz)
Fig11 Duisburg Nord – the park in fall (Source: Michael Latz)

图 12 北杜伊斯堡的卡波广场 (照片来源 : Peter Schäfer)
Fig12 Duisburg Nord Cowper Square (Source: Peter Schäfer)

town planner in partnership in 1968 (Fig.01-05). In the same year he started his academic career as lecturer at the Academy for Archi-tecture in Maastricht. 1973 – 1983 he was professor and chair of landscape architecture at the University Kassel, 1983 - 2008 professor and chair of landscape architecture and planning at the Technical University Munich. He has been lecturing and teaching worldwide, a. o. as guest professor at the GSD, Harvard University and as adjunct professor at the Graduate School of Design, University of Pennsylvania. The Technical University Munich has honoured his outstanding work as a researcher and a teacher in spring 2009 by bestowing the title "Emeritus of Excellence" on him.

Since the beginning of his office work and teaching, a main concern to Peter Latz has been a theoretically and scientifically based ecological urban renewal and herewith alternative environmental technologies like influencing of the urban climate by passive solar energy and greening of roofs and façades, rain water management, recycling. Since 1980 his work is focussing on the reuse of post industrial sites, their rehabilitation and cultural re-evaluation. He won numerous awards including the First European Prize for Landscape Architecture Rosa Barba 2000, the Grande Médaille d'Urbanisme of the Académie d'Architecture Paris 2001 and the EDRA Places Award 2005. With the award-winning project "Landscape Park Duisburg Nord", the metamorphosis of a 230 hectares industrial brownfield into a people's park and vivid part of the city he has gained world-wide reputation. It has been featured in numerous international publications and exhibitions including the Venice Architectural Biennale 1996, The CCC Barcelona's "Reconquest of Europe – Urban Public Space 1980 – 1999, several exhibitions at Harvard University between 1998 and 2003 and the MOMA's "Groundswell – Constructing the Contemporary Landscape", New York 2005 (Fig.06-15).Examples for projects being just in process are (Fig.16-20):

The "Parco Dora" inTorino, Italy: the transformation of a former Fiat car factory into a park near the city centre; The Park Ariel Sharon in Tel Aviv, Israel, being both a huge flood retention basin and a high standard recreation landscape and the Hiriya landfill rehabilitation, being the core of this landscape. The Avenue John F. Kennedy, Luxembourg: the transformation of the former motorway into an urban boulevard and lively public open space (Fig.21-23).

On occasion of Peter Latz' 70th birthday, renowned experts and colleagues from all over the world contributed with statements in a recently published booklet "Learning from Duisburg Nord" (ISBN 978-3-941370-07-4):

James Corner (Prof., School of Design, Univ. of Pennsylvania):……" Peter Latz seized the opportunity to work with the deep structures of the site. He did not impose his own will or signature, but instead cultivated the hidden processes and forces inherent to the site itself. It is a profoundly temporal work, essentially growing a new reality from an old foundation, a reality where traces of the old remain not simply as signs vestiges but as the alchemic medium within which the new springs forth. Latz is clearly an original Master, and Duisburg-Nord is his Masterpiece".

Christophe Girot (Prof., ETH Zürich, Switzerland):……"By letting the various contaminated surfaces develop slowly into fallow ecological time machines, Latz introduces a sustainable narrative dimension to a formerly vast and desolate site. And it is

彼得·拉茨1964年毕业于德国慕尼黑技术大学，其后的四年在德国亚琛工业大学研究所从事研究，并于都市计划工作室实习。1968年他以身为独立景观建筑师与城市规划师成立了自己的事务所，并与该研究所成为伙伴关系（图01-05）。同年他开始担任马斯特里赫特建筑领域的讲师，开始他的学术生涯，1973-1983年于卡塞尔大学担任教授兼景观建筑系主任；1983-2008年转任慕尼黑技术大学景观建筑系担任教授兼主任。他一直在世界各地进行讲座与教学，担任哈佛大学设计研究所客座教授，也在宾夕法尼亚大学设计大学部任副教授一职。德国慕尼黑技术大学在2009年春天为表彰他在教学与研究上的贡献，特别授予其荣誉教授之荣衔。

彼得·拉茨自从开始其实务和教学工作起，便关注于以理论与科学为基础的生态都市再生与替代性环境技术之议题，例如影响都市气候极大的被动型太阳能应用、屋顶与立面绿化、雨水管理与循环利用。1980年开始，他的作品主要聚焦于后工业区的再利用、再造与文化价值之重塑。他个人赢得了许多奖项，包括2000年欧洲景观建筑界的罗莎巴尔首奖，2001年巴黎建筑学会Grande M é daille D'URBANISME奖，以及2005年EDRA地区奖。得奖作品中之北杜伊斯堡景观公园(Duisburg Nord)系将230km^{2}的工业棕地转变为公众公园，建构都市

图13 Valdocco南端的公园和位于建物间约400米长的清澈雨水水道（照片来源：Andrea Serra)

Fig13 A Clear rain water channel 400 metres long mediates between the park and the adjacent buildings at the Southern edge of Valdocco (Source: Andrea Serra)

图14 秘密花园的初始景象（照片来源：Ornella Orlandini)

Fig 14 The "secret garden" at its very beginning… (Source: Ornella Orlandini)

precisely this combination of the industrial ruin with the potential of ecological rebirth that sets the precedent……But the most significant factor of acceptance in this park is its fundamental shift in use……We are witness to a significant change in society where natural and industrial elements are transfigured and are accepted mutually. All of a sudden the weight of the mining and steel history becomes incredibly ligjht and becomes the main cultural lever of a new pride. Inversely, the miracle of a blossoming "ecological" landscape rising like the phoenix from one of the most polluted sites on the planet, is both a surprise and an appeal to hope in our most troubled times, all to the honour of a landscape pioneer named Peter Latz."

Wolfgang Haber (Prof. em. Dr.Dr.h.c., WZW TUM):……"Thus there

中充满生命力的场域，因而获得世界级的名声，并在许多国际性展览展示及著文介绍，并分别在 1996 年威尼斯建筑双联展，巴塞隆纳的当代文化中心 (CCB)「再征服欧洲，1980–1999 年都市的公共空间」，1998–2003 年间在哈佛大学中多个展览，以及在纽约 2005 年「MOMA 的风潮 – 构建当代风景」展展出（图 06–15）。另外，目前正在进行中的计划包括意大利杜林的多拉公园 (PARCO Dora)，将城市中心的前飞亚特汽车工厂改造成一座公园（图 16–20）；以色列特拉维夫沙龙公园，这个公园既是一个大型的滞洪池，也是一个高级休闲景观区，其中赫利亚 (Hiriya) 垃圾掩埋场成为核心景观。在卢森堡的约翰肯尼迪大道乃是由先前的高速公路转换成林荫大道，并形塑成为有生命力的公共开放空间（图 21–23）。

彼得·拉茨 70 岁生日之际，来自世界各地的知名专家和他的同事于最近出版的小册子，名曰《向北杜依斯堡景观公园学习》（ISBN978-3-941370-07-4）载道：

詹姆斯·卡尔诺（美国宾夕法尼亚大学设计学院教授）：……“拉茨是一位具原创性的大师，杜伊斯堡是其个人的代表作。他抓住机会为基地打下深厚基础，他并没有对地区强加自己的意愿进行规划或落款，反而更加强调培养代代相传隐藏其中的在地过程与力量的展现。这是一个深奥且暂时性的工作，从旧的基础发展出新实质形式，其中不仅蕴含过往的痕迹，也是展现新的面貌的试金石基质”。

克里托福．格烈特（瑞士苏黎世大学教授）：……“彼得·拉茨将永续的叙事维度应用到先前广阔和荒凉的地区，让各种污染的地表，慢慢发展再次转入生态演替的时间机构中；他亦正确的应用生态再生

is created a diversity of activities and impressions which really gives a meaning to the in word “biodiversity” – due to the full and active integration of the people and their works.”

John Dixon Hunt (Prof. em., Dr., Univ. of Pennsylvania): “What a historian is tempted to learn from Landschaftspark Duisburg-Nord is how the past shapes, and is re-shaped by the present. And about usage (or reception) rather than design……Between the sublime and the beautiful, Peter Latz and Assoc. discovered a third term: not a dead and banal picturesque (the bane, anyway, of contemporary designers), but a mode of transfiguration that acknowledged both the extremes of experience within which modern people must conduct themselves. That this transformed landscape is also wonderfully photogenic is, I suppose, a “picturesque” bonus.”

Andreas Kipar (Dr., Lecturer Univ. of Genova and Milano,

图 15 杜林多拉未加盖的运河和步行街（照片来源：Latz + Partner）
Fig 15 Parco Dora – uncovered river + promenade (Source: Latz + Partner)

图 16 “未来的丛林”的最初场景，茂密的植相引人瑕思…，右侧是横跨钢铁林的 Passerella 和新的草坪公园，进而延伸到被称为”城市露台＂的 Mortara（照片来源：Ornella Orlandini）
Fig 16 The “futuristic jungle” at its very beginning, the lush vegetation still only to be imagined…, To the right the Passerella crossing the steel grove and the new meadow park, then linking to the “city balcony” of Mortara. (Source: Ornella Orlandini)

图 17 杜林多拉的梦幻花园（照片来源：Latz + Partner）
Fig 17 Parco Dora envisioned gardens (Source: Latz + Partner)

图 18 原生产大厅如今成为休闲活动场所（照片来源：Heidemarie Niemann）
Fig 18 The former production hall – now a fantastic setting for leisure and events (Source: Heidemarie Niemann)

图 19 杜林多拉通往城市的梦幻人行桥(照片来源:Latz + Partner)
Fig 19 Parco Dora envisioned footbridge to the city (Source: Latz + Partner)

的潜力将工业废墟作充分结合。…… 但最重要还是因为公园再利用本质上的改变，促使公园接受性佳。…… 我们都见证社会显著的改变，特别在自然与工业元素的转型与相互包容。他轻化采矿与冶钢的环境特征， 反倒是在主要文化层面上有傲人的新创意，一朵朵奇迹的“生态景观”花朵的绽放，就像凤凰飞出地球上最污染的土地，这不仅带来惊喜，也是在混沌时期中一种希望的展现 ，这所有的荣誉都归于彼得·拉茨”。

沃尔夫冈·哈伯（博士，荣誉博士，慕尼黑工业大学教授）：……“由于完全并且主动地将人融入他们的工作作充分的结合，因此可以创造出多样的活动及真实展现生物多样性此用词的意象”。

约翰·狄克生汉特（博士，宾夕法尼亚大学荣誉教授）：……“历史学家想要从北杜依斯堡景观公园学习的是，过去如何被塑造，又如何被现今再次型塑。它主要考虑的是利用性，而不是设计，…… 在崇高与美丽之间，彼得·拉茨和他的伙伴所提出的第三个专有名词「加值式风景如画般景观」：意指设计非常适合拍照的景观之思维，过去设计者经常设计风景如画的景致，并墨守此毁灭性思维，但此形式却无生命力而平庸，彼得·拉茨则将其转型为世所公认，现代人必须体验的极致经验"。

安德烈亚斯·卡帕（博士，土地 SRL 米兰热那亚和米兰大学讲师）：……“作为景观建筑师，我们传统上会满足于追求艺术的美丽意象，这也是一般公园与都市绿地所展现的景观特征，但是在社会成长及古典工业社会末期，这样的美学观点逐渐被抽离，也因此需要更多的创新过程来适应如此的转变，同时也重新定义建筑与栽培景观的

srl Milano): ……" As landscape architects we feed traditionally on the beautiful images of landscape art. This usually characterizes urban parks and urban green. The withdrawal from a society of growth and the end of the classical society of industry require more and more innovative processes of adaptation, which also re-determine our relation to the built and cultivated landscape……The current planning of the Parco Dora on a former brown-field in Torino……shows clearly, how visions can turn into general principles in a very short time."

Niall G. Kirkwood (Prof. Dr., GSD Harvard University): "Duisburg Nord is a wise, timely and significant project and set of design practices. Design proposes ideas in a landscape but it is through the craft of making and the medium of the constructed or reconstructed landscape that these ideas are projected as a material reality on a site,. This art is demonstrated throughout Duisburg Nird by the triple actions of "head, heart and hands"; with the head representing the analyzing, the structuring; through the heart, the passion, the human intensity of recovery and regeneration are demonstrated, through the hands the forming, the material. At Duisburg Nord it becomes an inventive, subtle and robust art form mirroring the intellectual activity of design, both as a daily form of practice and a s a personal aesthetic language……"

图 20 蓝厅 – 夜晚的神秘世界 (照片来源 : Fabrizio Zanelli)
Fig 20 The blue hall - a mysterious world at night. (Source: Fabrizio Zanelli)

关系。…… 目前在杜林 (Torino) 进行多拉公园的棕地再造计划 …… 就清楚的显示，远见如何可以在短时间内就转变成为一般性的准则 " 。

尼尔·柯克伍德 (博士，哈佛大学设计研究所教授) ：…… "北杜依斯堡景观公园是智慧、及时的重要个案与设计作品。设计乃是提出景观塑造的想法，但借由施工的技术与建造或重建景观，让设计概念被真实投射到基地之实物展现。北杜依斯堡景观公园以 "头到、手到、心到 " 三种面向来展现景观艺术，头到表示的是理性分析，而整体架构由心到与个人热情控制，人类的回复与再生之强度因之而被表现出来，手到可以借由型塑展现物质的特征。北杜依斯堡景观公园案已成为一创造性、微妙和强大的艺术形式，反映出设计的理智活动，它既展现经常性的操作形式，也展现个人的美学语言 …… " 。

马克·特莱 (美国加州大学柏克莱荣誉教授) ：…… "在北杜依斯堡公园，你很难感觉该从哪里进入公园，你身在公园何处或是该从何处进入其中，这是因为过去的经验法则在此处都无法应用，这个场域挑战我们过去对于景观型式使用的习惯及态度，但是也提供我们新的可能性。由于北杜依斯堡公园的设计、执行、成功的被接纳及被各界广泛的报导，已经成为如何处理荒废区域的国际级典范，就如同史卡帕日志成为历史架构中的博物馆典模。北杜依斯堡公园建立了一个新的领域，不仅只是在规模上，也在于其将工业废墟重塑其内涵，提供我们接近其中的老建筑，进而有更接近历史的机会。北杜依斯堡公园复苏之后，凸显本案并非由荒野处女地开发建造，而是透过工业废墟转换再利用，这个先驱性典范点亮了设计之路" 。

马丁·魏尔 (博士，Beracha 基金会理事长，以色列耶路撒冷博物馆

Marc Treib (Prof. em., Univ. Of California Berkeley):……"At Duisburg-Nord it is difficult to sense where to enter the park, to know where you are "in" the park, or to sense where to go inside it. That is because the historical rules no longer apply. The field challenges many of our old habits of use and attitudes about landscape forms, but it also offers us many new possibilities. As a result of its design, implementation, successful reception, and wide publication, Landschaftspark Duisburg-Nord has become the model for approaching spoiled sites internationally, as Scarpa's dialogues became the model for museums in historical structures. Duisburg-Nord staked out new territory not only in terms of scale but also in its re-contextualizing of industrial remains, providing access to its old buildings and ultimately, to its history. The many works that have followed in its wake look to this pioneer that blazed the path not through virgin wilderness, but through the industrial wasteland."

Martin Weyl (Dr., Dir. Beracha Foundation and Dir. Em. Israel Museum Jerusalem):……"At the same time Duisburg-Nord is also a very personal manifestation expressed in its design, through its understated civility, cultural knowledge, and profound intellectualism. In his design engagement in Israel, Peter Latz was able to adapt himself to a completely different climate, vegetation, and social contact by sensing and distinguishing the

图 21 施工中的原高速路 (照片来源 : Fonds d' Urbanisation et d' Amenagement du Plateau de Kirchberg)
Fig 21 Former motorway under construction (Source: Fonds d' Urbanisation et d' Amenagement du Plateau de Kirchberg)

图 22 更新后的 J.F. Kennedy 大道 (照片来源 : André Weisgerber)
Fig 22 Avenue J.F. Kennedy after transformation (Source: André Weisgerber)

图 23 J. F. Kennedy 大道（照片来源：André Weisgerber）
Fig 23 Avenue J. F. Kennedy (Source: André Weisgerber)

荣誉馆长）：……"北杜依斯堡公园设计者本质的投射，表现在设计、透过其底蕴的文化内涵、知识与高超的智能。彼得·拉茨在以色列的设计案中，通过感知并辨认出地区性的需要，把他自己的作品融入完全不同的气候、植被中并与社会连结。他创新、当代的个人风格，跨越传统的藩篱，并以精确、有创意的工程技术支持，提供我们创新如诗般的景观"。

俞孔坚（博士，北京大学建筑与景观设计学院院长，土人景观负责人）……"受到彼得·拉茨及其北杜依斯堡景观公园的启发，我在2000年设计的中山船坞公园是中国第一个大规模的工业废地再利用的景观设计案。个人受到他很深的影响，也铭感在心。……本公园的价值着根于一般大众、常见景观及环境伦理，已使此公园成呈现新的景观，主要整合国际与当代思维来营造文化自明性，也展现了新美学、新诗意的意境，呈现与过去传统景观截然不同的风貌"。

2012年2月24日下午2点整，在北京林业大学图书馆报告厅举办了一场与大师座谈会，与会者有国际景观园林师彼得·拉茨教授、台湾造园景观学会名誉理事长王小璘教授、北京林业大学园林学院王向荣教授、华中农业大学副校长高翅教授、清华大学农业景观学系系主任杨锐教授以及该系郑晓笛小姐。

王向荣：今天早上彼得·拉茨教授为我们进行了一场非常成功的演讲。在中国大陆，我们可以找到很多关于拉茨教授本人及其相关项目的出版品。我想，在座的各位也都对他很熟悉，所以我不再赘言介绍。这一位是台大教授王小璘，是我们的朋友，也曾莅临本校进行很棒的演讲，且曾多次来到中国进行学术交流。杨锐教授是清华大学景观学系系主任，高翅教授则是华中农业大学副校长，同时也是华中大学在

local essentials. Indeed his new, contemporary and personal style – transcending traditional boundaries and backed by precise and creative engineering – has given us a new poetry of landscape."

Kongjian Yu (Prof. Dr., GSLA Beijing Univ., Turenscape Beijing):…… "Personally inspired by Peter's thinking and his Duisburg-Nord Landscape Park, my design of the Zhongshan Shipyard Park, built in 2000, is one of the first large scale industrial site reuse projects in China; I was influenced by him and I am thankful for that…… The values rooted in common people and common landscapes, as well as an environmental ethic, have made this park a new landscape with cultural identity integrated into an international and contemporary way of thinking which has demonstrated that

对话现场

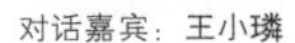
对话嘉宾：王小璘

对话嘉宾：高翅

对话嘉宾：王向荣

对话嘉宾：杨锐

翻译：郑晓笛

景观建筑领域中的领导者，此外，我们今天下午的口译人员是郑晓笛小姐，目前是清华大学的学生兼老师。我是下午座谈会的主持人王向荣，在北京林业大学园林学院教书。今天下午我们将有两个小时的时间进行对话和交流，在对话后，如果各位有任何问题或意见，可直接发言。现在我们先将麦克风交给高翅。

高翅（以下简称「高」）：我很高兴能有机会参加此次研讨会。我认为要成为一位优秀的景观设计师，最重要的特质之一，就是不断处理与设计师、其他人和大自然之间的关系，或更广泛地说，是处理人类和文化。我认为拉茨先生重建工业用地，反映出他对人类及人类文明的尊重。我的观点是工业文明遗产的重新配置，不仅是我们景观设计师的责任，也是城市规划师和建筑师的责任，因为它是在更广大之文化下产生的产物。

在与人类共存时，每一个地点、每一块土地都有其本身的记忆，因此，我们不应该为了创造新事物而抹净旧记忆。我们创造的东西，必须来自人类行为与当地自然的过程，这是我个人对今天早上进行之演讲的理解。

谈到彼得·拉茨先生关注的土地类型，我偏好以「坏地方」一词取代「劣质土地」。事实上，我们在将一个旧地方转变成一个新地方时，必须对自然、对人们、对历史、对文化以及对土地，都怀抱着尊重和虔诚的信念，因此，我们创造的风景应与中华民族老祖先的理想和谐以对。要做到这一点，我们就必须找到一种内在的和谐，谢谢。

彼得·拉茨（以下简称「拉茨」）：高校长的评论虽短，却非常有见地，我认为他下了一个非常完美的结论。

高：我有一个问题想要请教您。您展示了很多所谓的「坏地方」，我的问题是，您选择坏地方的标准为何？

拉茨：坏地方大约是150年前之工业时代开始时遗留下来的问题。当时，已在许多领域中决定一个地方是否会发展成坏地方了，这些地方如果一直保持这种状态，即无法再作为生活环境，我可能也无法取得任何可定义坏地方的标准，这一切都取决于人们如何在那个地方生活，如何处理有问题的土地。

高：如果有两个选项供您选择，一个是纺织厂，另一个是重工业厂房，您会选择哪一个选项？

拉茨：从来没有人提出此问题。如果我真的可以选择，我会选高炉厂，因为身为景观设计师，我不认为纺织厂的周边有可供发挥的空间。当然，地方的历史、风气将是很重要的标准。

王小璘：听完拉茨先生的演讲，使我了解了一个成功的景观设计师应具备的特质。是什么特质呢？首先，是敏锐的洞察力。在他的作品中，我发现他能看见别人无法察觉的事物，例如，他可以发现50年前被弃置的土地有什么问题，然后进一步发展出新的概念，这就是他敏锐的洞察力。第二个特质是坚持。在拉茨先生的作品中，他坚持将

the meanintg of "The new Aesthetics" and "The new Poetics" differ dramatically from those which are considered Classical Landscapes."

2012/2/24. 14:00. In Auditorium room of Beijing Forestry University held a forum of landscape masters. The participants are Professor Peter latz, Professor Xiaolin Wang, Honour Principal of Chinese Landscape Architect Association, Professor in Taiwan Xiangrong Wang from Beijing Forestry University, Professor Kaochi, Vice President of Huazhong Agricultural University, Professor Rui Yang , Head of the Department of Landscape Architecture, Tsinghua University and Miss Xiaodi Zheng, a teacher from Tsinghua University.

Xiangrong Wang: Peter Latz gave us a very successful lecture this morning. In China there are a lot of publications about Professor Peter Latz and his projects. I think everybody here knows a lot of information about him, so I think I don't need to introduce more about him. This is Xiaolin Wang, a chair-professor of Landscape and Urban Design in Chaoyang University of Science and Technology in Taiwan. She is also our friend and has given a successful lecture here and has visited the mainland many times for academic exchange. Mr. Yangrui, the head of the Department of Landscape Architecture in Tsinghua University. Professor Chi Gao, the Vice principal of Huazhong Agricultural University. He is also the leading professor in the field of landscape architecture in the university. And our interpreter this afternoon is Xiaodi Zheng , she is a campus student and a teacher in Tsinghua University. I'm the host this afternoon, Xiangrong Wang , a professor in the School of Landscape Architecture of Beijing Forestry University. We have two hours for dialog and conversations this afternoon. Meanwhile, after the dialogue, if the audience has questions, you may ask or express your opinions directly. Let's give the microphone to Chi Gao .

Chi Gao (Name as 'Gao' below): I'm glad to have this opportunity to attend the seminar. I think to be an outstanding landscape designer, one of the important characters we need to possess is to continually deal with the relationships between designers, other people and nature, or in a wilder extent, with human beings and culture. I think Mr. Latz's reconstruction of industrial land reflects his respect for human beings as well as human civilization. In my opinion, the reconfiguration of the heritage of industrial civilization comes not only from our landscape designer but also from city planners and architects because it is product of our broader culture.

As with human beings, every site, every piece of land has its own memory. For that reason, we should not create new things

坏地方完全转型成新风景，我认为所有的年轻景观设计师都可以向他学习。

还有一个例子：台湾民众正在为林书豪疯狂，我认为林也向年轻一代示范了坚持的典范。他在高大的西方人之间打篮球、完成他的学业，都是经过不断的努力和自我成长，才能取得今天的成就，由此可知，坚持一件事，最终即会促成成功。我想请教拉茨先生一个问题：在您的背景中，哪一部分促成您拥有这样的特质?

拉茨：这是一个好问题，但是，我可能没有很好的答案可以回答您。当我小的时候，大部分东西都被战争摧毁了，从那时我就知道，我们必须保存和尝试利用所有的东西。另外一点是，我曾经与许多矿工和工人接触，因此了解资产阶级与工人之间的差异，中产阶级想到的很多东西，对劳动阶级而言没有太大的意义，因为对工人而言，工作才是他们的生活。

我希望小组讨论能探讨此主题的两面。我们必须考虑改进现有工业建筑的质量，尤其是经过去年的日本海啸和核漏灾害后，我希望新的工业用地不会再出现同样的问题。

其实景观设计师在这方面能做得有限，但是，我们必须处理城市基础设施，而这些设施通常不是为人们设计，我想以大约2000年前的街道开发和利用作为很棒的小范例。这与今日的基础设施有何关联呢?我们会警告孩子们不要跑到街上，因为实在太危险了，但是我们应该为人创造空间，而不是为机器。

我们在大学中，最重要的是与其他专业沟通，例如与交通工程师沟通。对于景观设计师而言，种树很容易，而了解其他行业却很困难，但是为了带领团队，学会与其他专业合作也相当重要。

杨锐（以下简称「杨」）：大家午安，我很高兴能有机会在此与各位聊聊，不幸的是，我是这里唯一在过去20年间没有参与坏地方重建工作的人，而晓笛总说她很幸运，因为她现在正在从事相关工作。我曾经参与自然和文化遗产的保护工作，所以经常造访九寨沟等风景名胜区，部分联合国组织的会议也会在这些自然遗产地举行，所以我看到这些美丽地方的机会很多。目前晓笛在清华大学攻读博士学位，她在选择研究主题时曾计划研究世界遗产，但是我给她的主题是坏地方的重建工作。

我曾经写过一篇文章，谈到从传统园林设计到现代园林建筑曾经历三个面向的转型。第一个变化是从仅服务少数人转型为服务广大公众，我们可以清楚地看到此过程。第二个变化是从小规模或单一规模，转型为较大且多元规模的公共公园、较大的国家公园系统、绿园道等等，今时今日我们也可以看到此变化。第三个变化是从单纯的视觉艺术转型为多样化的学科，我们可以了解拉茨先生演讲中的「新美学」，当然不仅是视觉，更涵括了广泛的园林建筑，其中还包含了社会、政治和经济的议题。

演讲结束后，我想要谈一谈从古典园林到现代景观的变化，并加入两项我本身的理解。首先是从孤立的美丽地方转型为整体的尺度，我们将以地球、国家、领土、地区、城镇、村庄和公园为尺度，不再局限于对地区范围的理解。我们如何才能将对尺度的新理解实践于社会中?其次，哪里可以指引我们的新研究?工业废弃地的转型相对强烈，因此从拉茨先生的思想和作品中可以看出一些东西。

我想要请教他的问题是，已开发国家和发展中国家该各自如何看待重建工作?第二个问题是，不同的文化背景产生不同的情况，是否意味着在美国、欧洲和中国进行重建工作时，每一个区域都会有鲜明的特色吗?我想要了解，在德国，重建工作的特色为何?您可以给中国一些建议吗?

拉茨：几个月前我在访问北京时，参加了一个会议，会议举行地点原本是一座炼炭厂，我在会议结束后发现厂内的技术与设计情况与德国完全一样，但在社会性方面却相当不同。随着全球化的发展，分享技术知识变得更容易，但是，最重要的是要找出中国和德国之间的

at the cost of wiping clean the old memory. The things we create must come from human works and natural processes on the site, this is my personal understandings of this morning's lecture.

Refering to the type of land that Mr. Peter Latz works with, I prefer to use the term 'bad place' as opposed to 'inferior land'. In fact, when we transform an old place to a new one, we must be respectful, and devout, in our reverence, to nature, people, history, culture and the land. Thus the scenery one creates must be a harmonious with the ideals of our Chinese ancestors. To do so, one must find a kind of inner harmony. Thank you.

Peter Latz (Name as 'Latz'below): Although Principal Gao's comment was very short, it was very reasonable. I think it is a very perfect conclusion.

Gao: A question I would like to ask you, is that you show many places which you called 'bad places'. My question is, what is your criteria for the selection of bad places?

Latz: Bad places are the legacies of the industrial age which started around 150 years ago. Already at that time it was decided in many areas if a place would develop into a bad one.

These are the places which cannot be used as a living environment if they would stay as they are.

I have no defined criteria for a choice I'm afraid. It all depends on how people are living there, how they are dealing with the place in question.

Gao: I mean there is a selection for you to choose from, one is a textile mill, the other is a heavy industrial plant, two projects you can choose from, which one will you choose?

Latz: This question has never been posed to me. If I really could choose, I would take the blast furnace plant, because as a landscape architect I don't see much room for activities around a textile mill. Certainly the history of the place, the genius loci, would be an essential criterion.

Xiaolin Wang: From the experience of Mr. Latz's lecture, I have learned about the character of a successful landscape architect. What are these characteristics? The first is sensitive insight. In his works, I find that he can see and visualize things which others don't realize. For example, he can find problems in land abandoned 50years ago and then further develop a

彼得·拉茨报告会现场

相似性和差异性，虽然我不懂中文，却依然可利用我的眼睛解读周围的环境，并找出地方特色。

我们可以理所当然地想，一个景观设计师不可能改变一个风景地的人口密度。当老工业城市中的产业不再进行生产时，人口数就会下降，城市也会萎缩，这是景观设计师不能改变的事实，而他可以做的是以完整的景观刺激城市再生。

高：我有一个问题想要请教主持人。我相信王教授曾看过一些彼得·拉茨的景观作品，我想要知道您在造访过他的许多景观作品后，有何感受？谢谢。

王向荣：这是一个很难回答的问题！我曾在欧洲看过他今天谈到的项目，也曾在他的书中读到过。我想要保留高教授的问题，将相关讨论交给拉茨先生，因为我无法找出一个很好的答案。

但是我认为，拉茨先生能在全球享有盛誉的原因，就在于他可以将一个地方的历史和生态、艺术紧密结合。从历史的角度看，也许在20年或30年后，有很多人会关注生态的问题，但是有很多人仍将美学与生态视为对立状态，当一项设计较关切生态时，就往往会减少对美学元素的关注，也会降低用户需要的实用功能。

我个人的观点是，拉茨先生可以如此吸引大众的目光，最重要的原因之一就是他能将艺术与生态结合。艺术与生态之间一点也不矛盾，今日的许多自然景观就是文化的代表，我相信他有他自己的考虑，并将这些考虑带入他的作品中，而使他的设计如此与众不同。我们认识世界上或德国的很多景观设计师，但是拉茨先生与别人不同，因为他有自己的想法，如果你没有自己的想法，就只能使用其他设计师的方式设计，这是我非常欣赏他的原因。

将设计带进实际工作中，事实上两者间存有一段距离，因为很多人都有自己的想法，却没有机会实现。我认为，拉茨先生备受瞩目的第二个原因是他具备很多技能，能让他轻松将设计融入作品中。他不仅在各个领域展现其科学技术，且更进一步地发展新技术，我听说他做了很多与太阳能建筑、绿色建筑、屋顶绿化、雨水收集、材料再利用等有关的研究。他进行这些工作很多年了，所以有时间和经验可以进行实验，再将结果转化至实际的工作中，他依据自己的调查获得许多与植物有关的知识。他不仅在学校，甚至在家里进行众多的研究，今天早上他展示的最后一张照片就是他的住家，他为这一幢房子做了一些工作，我相信，这也为他带来了成功。所以，我认为一位景观设计师应该具备两项特质，一是思想，另一则是能力，唯有拥有能力，拥有每一面向的深刻知识，才能实现你的想法，否则，可能只能成为一个不断重复他人经验的平凡设计师。我常常建议我的学生多思考，并多进行实际的工作。

高教授的问题非常好，我确实看过他的许多作品，之前也曾带领许多老师和学生到德国参访，而且去过很多地方，旅行后，许多学生都说这是一趟历史遗迹之旅。当我们第一次看到作品时，真的感到很震惊，但是多看一些他的作品后，就逐渐变得较习惯了，当时有一位非景观设计领域的导游觉得我们很奇怪，他不了解我们为何能从此类型的景致中得到如此大的享受。

在评论一项设计时，讨论好不好、美不美、能否吸引你或为你带来视觉享受，都已是过时的说法，比这些评论更重要的是它是否能保存土地的景观与记忆，这才是我们工作的重点。他完成的作品不在于创造美丽的景观，虽然这也是工作的一部分，而是在更新生态环境、带进历史，然后将快乐带给当地居民，并将珍贵的东西留给我们的后代。这是最重要的事。

我也有一个问题。导游不懂为何我们喜欢这些风景，他认为一般人不会选择去那里，而这些一般游客也不在我们的领域内。大多数中国人的价值观都很雷同，皆未深入了解一个地方的历史价值以及生态复育的意义，因此，我认为在中国，他的部分作品可能无法被接受或理解。

我看到的一件作品，名为「Buergerpark Hafeninsel」，我真的很

new conceptualization of it. This is due to his sensitive insight. The second characteristic is persistence. In Mr. Latz's projects, he persists in the complete transformation of a bad place into new scenery. I think all young landscape architects can draw lessons from him.

As another example, the public in Taiwan is crazy about Jeremy Lin. I think he is also presents an example of persistence for the young generation. He continued to play basketball among tall western people while completing his studies. After continuous effort and self-growth, he achieved fulfillment today. So persisting at one thing will eventually lead to success. I would like to ask Mr. Latz a question, what part of your background led you to develop such a character.

Latz: I think it is a good question, but maybe I don't have a good answer. When I was a young child, most things were destroyed because it was war. But I know that at that time we had to keep everything and try to use everything. The other thing is that I was in contact with many miners and workmen and I knew the difference between the bourgeoisie and the workers. The middle class thinks about a lot of things that does not mean much to the working class. For the workers, there work is their life.

I hope the panel discussion addresses the two sides of this subject. We need to consider improving the quality of existing industrial constructions. After last year's tsunami and nuclear leak in Japan, I hope that new industrial sites will not have such problems.

Actually, as landscape architects we cannot do much in this respect, but we have to deal with urban infrastructure, which is usually not designed for the people. As a wonderful small example from our history I would like to mention the development and use of streets some two thousand years ago. And how do we relate to this infrastructure today? We warn our children not to go on the street because it is too dangerous. We have to create spaces for people not for machines.

We are in a university. The most important thing is to communicate with other professions, with traffic engineers for example. For a landscape architect, planting trees is easy and understanding other professions is difficult. But it is crucial to learn to cooperate with other professions, in order to be able to lead a team.

Rui Yang (Name as Yang below): Good afternoon everyone, I'm very glad to have this opportunity to talk with everybody here. Unfortunately, I'm the only person here who hasn't been involved in the reconstruction work of bad places in the last 20 years. However Xiao Di always says she is less lucky because she works on them now. I used to be involved with the protection of natural and cultural heritage sites, so I could often visit scenic spots like Jiuzhaigou and so on. Some meetings with the United Nations organization were also held at these natural heritage sites, so I often got the chance to see these beautiful places. Xiaodi is pursuing her PhD degree in Tsinghua University now. When she chose her topic of research, she intended to study world heritage, but I assigned her this subject of the reconstruction work of bad place.

I wrote in an essay that there are three main aspects to the transformation from traditional garden design to modern

欣赏那一座公园，是 20 年前建造的公园。我不知道那一段期间德国的情况如何，如果是 1999 年的德国，会接受他的作品吗？如果不能，他有方法说服公众吗？为其作品取得成果也是非常重要的事。许多作品无法进行，不是因为我们的设计师没有能力或尚未觉醒，而是因为此作品无法被接受，所以在进行的过程中会出现很多障碍。我的问题是，他的设计在 20 年前能被接受吗？如果不能，他会怎么做？

拉茨：您的看法有 90%，我都同意，现在我必须谈一些细节部分。国际建筑展（IBA）始于 1989 年，结束于 1999 年。在 1989 年之前的 20 年，此作品完全不会被接受，没有人会在已经开始走下坡的行业中出现。IBA 的原则是鼓励和支持实验性作品，甚至在我们开始设计之前，周边的居民就已参与此过程，他们针对如何使用这一座公园提出了自己的想法和建议，因此，作品的接受度一直很高。

在第一个例子中，这种类型的公园是为住在 15 至 20km 范围内的居民而建造，最后却演变成吸引观光客的景点。

在德国，有 60% 至 90% 的人会认为葡萄园别墅是理想的景观，我本身则很热爱意大利花园，但是在我们的文化中，工业生产是很重要的一部分，我们不能把它藏起来，做一块绿色地毯覆盖住一切，我们必须处理它，也处理污染。

王小璘：从您对作品的说法，显示当地民众是主要的使用者，但是随着时间的推移，也许老一辈将消失，开始由下一代在那里生活，您期盼下一代如何看待您的作品呢？或是您对年轻一辈看待此类型新景观的主要期待为何？万一创作的过程和历史未经传承下去，在未来 10 年，仅被视为一个美丽的景观时，你建议年轻的景观设计师如何处理此问题？这算是一个问题吗？

拉茨：我很惊讶年轻一代会是第一个进入这个地方的族群，12 到 17 岁的女孩男孩会在那里真的会令人很吃惊。对他们而言，我们在原来为锰矿厂的地方，创造了一个 15000m^2 的特殊空间，我们改造了 6 到 7 块区域，让他们可以运动、游戏，或单纯只是坐在那里与其他青少年聊天。年轻人从河的另一边来到这里见他们的朋友，除了许多小的「游戏点」外，我们还提供了两个大型儿童游乐场，也吸引着人们愿意从很远的地方前来，因此，我敢肯定下一代应该会喜欢这里。

在如何处理后工业景观方面，在我的国家，年轻的景观设计师们都会受到很好的训练－甚至是在硕士课程中。

王向荣：在座的各位现在可以发问，中文或英文都可以。

发问者 1：如同您在以色列的作品，当地的土地有甲烷，而甲烷易燃、会引起爆炸，也是温室气体，我的问题是，您是否有方法重复使用或处置垃圾填埋场的甲烷？

拉茨：当然，可以使用一个密封系统，将沼气收集后，送到 2km 外的工厂做为热源使用。

发问者 1：依此看来，在过程中必须先处理填埋场的废物，再于现存地点建造新的景观。我想知道您对于处置废物，并还给土地原始风光有何建议？或者是否有一些技术或方法可以协助做到这一点？

拉茨：正如您的想象，将一座高 65m，包含家庭废物的山，移到另一个地点是不可能的事，此外，您将这一座山移到哪里，哪里就会留下垃圾。

高：我懂他的问题，他想知道是否有其他替代方法处置废物，譬如焚烧？

拉茨：此类型的废物很少使用焚烧法，因为燃烧过程需要时间，

观众提问 1

landscape architecture. The first change is the transformation from serving only a few people to the broader public, which is a process we can clearly see. The second change is the transformation from small scale, or single scale to the larger multi-scaled public parks, larger national park system, greenways and so on, which we can also see these days. The third change is the transformation from a pure visual art to a multifarious discipline. We can understand that the 'new aesthetic' in Mr. latz's lecture certainly refers not only to the visual, but also to the broader context of landscape architecture which contains social, political and economic concerns.

After the lecture I wanted to talk about the change from classical garden to modern landscape and add two of my own understandings. The first is the change from isolated beautiful places to a holistic scale. We take the earth, state, territory, region, town, village and park as scales. We no longer limit our understanding to the range of region. How can we join this new understanding of scale with social practice? The second, is where to direct our new research? It is relatively strong in the transformation of industrial wastelands. So something can be seen from the thought and work of Mr. Latz.

The question I want to ask him is how should we approach reconstruction work differently in the developed and developing worlds? The second question is that different cultural backgrounds produce different situations. Does that means that reconstruction work in America, Europe and China has its own distinct characteristics in each region? I want to know in Germany, what are the characteristics of reconstruction work, and is there some advice you could give to China?

Latz: During my last visit in Beijing some months ago, I participated in a meeting which took place in a former coking plant. After the meeting I found that in fact the technological and design situation is exactly the same as in Germany, but the social aspects are different. With globalization, our ability to share technical knowledge becomes easier. But the most important thing is to identify both the similarities and the differences between China and Germany. Although I do not understand the Chinese language, I can still use my eyes to read the surrounding environment and identify the characteristics of a site.

We can take for granted that for a landscape architect it won't be possible to change the population density of a

而且即使在干燥的天气下，垃圾含有的水分也很高，实际上可能比预期还要高很多，所以焚烧法通常行不通。另一方面，将会有一座新的回收厂建在山脚下，那里将回收大量的垃圾，部分流程即与焚烧技术有关。

发问者2：我是《中国环境报》的记者，想要请教您一些问题。第一个问题是，这是您第一次来中国吗？早上的演讲中，您曾提到非常喜欢中国的景观规划和城市建设，所以，您是否可以针对这些主题发表一下看法？有没有一些案例可以评论指教？

拉茨：中国的景观至少在两个方面影响了欧洲的景观规划。第一，在过去的150到300年间，中国对欧洲园林的理论有极大的影响，特别是英式风格的花园，它们深受中国园林的照片和绘画影响。第二是与一项事实有关，即是中国植物物种的多样性是欧洲的3倍，引进中国植物对于欧洲花园和公园使用的植物有极大的影响。

当您亲自来到中国，一方面会对高质量的景观和城市规划感到印象深刻，但另一方面也会被惊人的建设与拆解速度震撼，您看到的将会与书本或媒体讲述的完全不同。我第一次来中国时，住在一个历史悠久的两层楼旅馆，那一栋楼的外观和气氛散发出我从文学作品中读到的感觉，但是现在这一座建筑已不存在了。我想，欧洲进行项目的速度，最多只有中国的十分之一。

发问者3：我想要请教拉茨先生一个问题。您在创作每一项作品时，本身必然会有某种期许，请问您如何掌控作品以达到目标？如同您提到的煤矿作品，您想要保留矿区原来的样子，并让当地保有工业记忆，但是，最后旧设备仍被青少年破坏，或许是因为他们与这一段记忆没有关联。您也说，青少年是这一座公园的主要使用者，请问您如何应对这些预期的使用者呢？关于此问题，您是否有任何可分享的意见？

拉茨：请先容许我澄清一些误解：我指的是高炉厂改造，而非矿区改造，在高速公路的另一边有一座旧矿区，矿井已经关闭，还有一个不相关的大焦油湖也封闭了，我没有谈到这一块鲜少被使用的区域，而是将它留给大自然演替。

有一些人为破坏，但只发生在刚开始之时，正如刚才的解释，年轻人特别喜欢这一项工业遗迹。青少年不是公园的主要使用者，而是成为他们领域中主要且唯一的用户。

译者：您想要问，什么人是公园的真正使用者吗？

发问者3：我想问的是，您能掌控预期结果的程度为何？如同儿童游乐场是专为儿童设置，矿区项目是为了保留城市的记忆，然而，最终则由青少年使用这一座公园，这与原来的想法是否有冲突？

拉茨：这个地方是为所有人而设计，可以在任何时间使用，一天24小时，一年365天，它不是只为一群人而设置。

王向荣：您是一位设计师，您的太太也是。您是否经常与家人讨论您的工作？您如何在工作与家庭之间找到平衡点？因为当家庭与工作混在一起时，最终将会变成整天都在工作，而令人感到十分讨厌。您和您的太太对于工作与家庭生活混淆，是否有任何意见相左之处？

拉茨：这是一个非常好的问题。我们永远无法找到最理想的平衡点，这是我们专业人士很典型的特色。身为设计师，大部分的时间都在想自己的作品，也就是说，您不会将笔放下，表示现在工作已做完了，即使在度假，也无法真正把家庭生活与工作分开，但是，我与我的太太、儿子可以很有效率地一起工作，我认为在许多方面都是很好的一件事。

王向荣：我也想要请教王教授这个问题。

观众提问2

landscape. When an industry ceases manufacturing in an old industrial city the population also declines, the city shrinks. This is what a landscape architect cannot change. What he can do is to stimulate the city's regeneration with an intact landscape.

Gao: I have a question for our host. I believe professor Wang has seen some of Peter Latz's landscape work. I wonder after visiting many of his landscape sites what are your feelings? Thank you.

Xiangrong Wang: It is a difficult question ! The projects he talked about today I have seen in Europe and read about in his books. I want to set aside Professor Gao's question and leave the discussion to Mr. Latz because I don't have a good answer.

I think though, the reason why Mr. Latz has such a reputation in the world is that he can bring a place's history closely together with ecology and art. We can see this from a historical perspective, maybe after 20 or 30 years, but at that time, there were a lot of people concern about the ecological problem. To the question of aesthetics, lots of people regard it as in opposition to ecology. When a design is more concerned with the ecological aspects, it tends to have less concern for the aesthetic factors and the practical functions for the users.

In my personal perspective, one of the important reason he can attract the public's attention is that he can combine art with ecology. Art and ecology are not contradictory. Much natural scenery today is the representation of culture. I believe he has his own considerations and brings these into his work, which makes his designs different from others. We know many landscape designers in the world or the designers in Germany. But Mr. Latz is quite different from others because he has his own considerations. If you don't have your own ideas then you can only do the designs in other designers' way. This is the reason why I appreciate him very much.

The second one is that there are gaps to bring your design into real work, many people have their ideas but have no opportunity to achieve them. I think Mr. Latz has many technical skills to bring to a project. He not only presents scientific techniques in various fields but also advances new techniques. I heard he has done much research on solar buildings, green buildings, roof greening, rainwater collection, material reuse and so on. He has done these jobs for many years, so he has had the time and experience to

王小璘：我简单地回答这个问题。我的先生是一位教授，也是一位建筑师，我们既做建筑，也做景观，而且经常意见分歧，我的答案很简单，最重要的是积极面对问题。多年以来，我们已学会从相反的观点争论，再逐渐达成一些共识，景观与建筑通常必须整合在一起，不能分开处理。景观与环境有关，建筑则是指单一或一群建物，建物可创造出实体空间，因此，景观和建筑就如同镜子的两面，但是，景观设计师的空间概念不如建筑师那么强烈，这是训练造成的差异，不过，景观设计师必须以全球规模进行思考。若是从这个角度来看，我是从大到小来看事情，我的先生则是从小到大着眼，正因为如此，我善于运用逻辑处理大问题，而他较具有创意，所以我们一个人用左脑，另一个人用右脑，最后，结合成一体。

所以，我想要建议每一位设计师，不仅应学习以客观、有逻辑的方式分析和处理事情，也应该学会培养自己的创造力，因为您在工作过程中无可避免地需要它。传统的建筑教育总是不断地强调个人风格，而造就了不同的学院，或许不同的专业可提供互补的作用，不过坏处是争论不休。当我们的孩子尚年幼时，不懂为什么我们总是在争论，直到他们长大后就了解了，现在，我们的家庭旅游就好像是在进行专业的调查行程，我们看景观、看建筑，看所有的东西，我们终于成为一体。

发问者4：我想要提出一个问题。在我们周遭可以发现很多严重污染的地区遭到弃置，这些区域可能会为当地居民带来极大的伤害，所以，我想请您针对消除污染和为当地居民及野生动物创造安全的生活环境方面，提供建议。

拉茨：您的问题已点出这些作品最需要讨论的面向，不过，我认我们无法提供一个统一且一致的答案，因为每一个地方都很不同，例如，制药厂可能会有一些特殊的污染，所以必须在每一个地方都进行仔细的研究，以了解具体的情况。不过有一点很清楚就是在90%的情况下，这些地方都不能开发成住宅用地。

发问者5：但是，这种情况却发生在今天的中国，这些遭到污染的土地已被开发成住宅区。

译者：出现这种情况，我们如何能获得人们的信赖？

发问者5：没错。

拉茨：这真是一个大问题，只有当地才能回答，而且必须仰赖人们的经验和知识。

王向荣：同学提出的问题非常好，表现出对人类的关注。太严重与困难的问题，恐怕无法在此轻易地解答，让我改变一下问题，轻松一下。现在，我有一个原则，就是在家时不谈论工作，因为我们会互相争论，导致无法达成协议，在办公室则无妨，因为在办公室中争论是很正常的事。

我的问题是，他的家同时也是他的办公室，所以可能显得更复杂，我也不清楚，我的许多朋友和同事都有类似的家庭，我可以向他们学习一些经验。

王小璘：如同王教授所言，这种情况需要坚持和宽恕，有时候，如果只是一味地坚持，您将无法妥善地处理问题。

杨：我现在正在翻译拉茨先生的一本书，因此阅读了其中一些章节，学到了很多东西。我认为景观设计真的可以将破烂变成珍稀的天物，但是另一方面，确实也会让神奇幻灭，一些美丽的景点和自然遗迹就是实例。所以，我认为关键在于我们如何行动，如何以规划和设计传达景观建筑的魅力。您可以从他今早的演讲及一些著作中了解，透过他的设计，可以将一个冷冰冰的工业建筑变成一个温暖的地方，使这

experiment and then put it into actual work. His knowledge of plants could not be achieved without his own investigation. He has done much research in his school, even in his home. The last picture he showed this morning was of his house on which he has worked and I believe that this is what has brought him his success. So I think to be a landscape designer, you should have two characters, one is your own character and the other is your ability. Only with ability, with a deep knowledge of every aspect, can you carry out your thoughts. Otherwise, you may be an ordinary designer always repeating others' experiences. I often suggest to my students to think more and do more practical work.

Professor Gao's question is very good, I actually have seen many of his projects. I have brought many teachers and students to Germany before. We traveled to many places. After traveling, many students said it was a tour of historical ruins. When we saw this project, at first glance we were really shocked but after seeing more of this work we became more accustomed to it. A tourist guide who was not in our major could not understand us and wondered how we could get so much enjoyment from this type of scenery.

To comment on a design, whether it is good, beautiful or can it attract you, can it bring you visual pleasure is a outdate topic. What is more important than these comments is whether it can remain a land's scenery and its memory. It should be the key point of our work. The projects he has done were not to create great scenery, although that is part of the work, but to renew the ecology, to bring out the history, then bring happiness to local residents and keep the valuable things for our descendents. This is the most important thing.

I have also a question. The tourist guide doubted that we enjoyed these landscapes. He thought common people would not choose to go there. Those common visitors are not in our field. Most Chinese people have common values and they don't have a deep understanding about the historical value of a place as well as the meaning of ecological restoration. So I believe some of his works might not be accepted or understood in China.

One of the projects I saw was Buergerpark Hafeninsel. I really appreciate that park but it was constructed 20 years ago. I don't know the situation in Germany in that period. If it was 1999 in Germany, would his project be accepted? If it could not be accepted, does he have some methods to persuade the public? That he brought this project to fruition is also very important. Many projects are not constructed not because our designers don't have the ability or awareness but because the project could not be accepted. So there are many difficulties in the process. My question is would his design be accepted 20 years ago? What would he do if his design was not accepted?

Latz: I agree with 90 percent of your perceptions.

I must go a bit into details now. The International Building Exhibition (IBA) started in 1989 and ended in 1999. 20 years before 1989 this project would not have been accepted at all. Nobody would have shown the already beginning decline of the industry. The principle of the IBA was to encourage and support experimental projects. The residents of the surrounding quarters participated in the process even before we started with the design. They contributed with their own ideas and suggestions how to use the park. Therefore the acceptance has always been very high.

些作品似乎也能拥有它自己的喜怒哀乐。借由植栽设计等，可以将一个含有冰冷死寂之金属的地方变成活泼充满活力的场域，如今我们已可在许多案例中看到这样的精神，我真的认为这就是景观设计的意涵，或是换言之，这也是这一门学科的魅力所在。

我的问题是，他如何将这种精神和情感元素带进工业用地的规划设计？我曾经与他接触，从外表来看，他是一个严肃的德国人，但是我能在他的一些作品中，感觉到他内在情感的流露，而这些情绪都是发自他的内心。这是我想要问的问题。

拉茨：我当然不希望毫无情感，不过，我认为最关键的任务，是找出并发展对作品而言至关重要的架构，否则就不可能保有您或其他人带进来之物体的控制权。尤其是，如果进行的是一项大型项目，扎实的架构能让您决定选择的元素及安排的规则，并可将不同外观的物体完美地整合在这个相当中性的架构中。

未来的用户和造访者可以决定运用的元素。有一句话听起来很像笑话：「摄影师可决定拍摄对象漂亮与否。」但是我必须明确地指出一点，那就是维护，对整个发展过程而言非常重要。例如，当您修剪一棵树，即已违反自然架构，而语义的质量或表达方式也会改变这个空间。

维护工作是指能控制其他事物也许无法改变的东西。它可以确保您希望人们在未来的岁月中看到的状态与空间。缺少维护，地方会失去大量的讯息和元素。

景观建筑师从来不会计划这个地方一年后的样貌，我们通常是描述 15 年，甚至更长时间之后的状态，此亦表示我们描绘的是长成的树木。您无法避免自然灾害或其他因素改变您的设计，所以尝试完全控制一切，是不可能的任务。

发问者 6：我曾经在建设部工作超过 30 年，并一直在这个领域中，许多我曾接触过的设计师都对他们的工作感到困惑，我认为此问题可能与国家的制度有关。我国的制度与其他国家截然不同，我们的设计师通常会依据当地的情形、在地文化、历史、环境和社会背景，创作出一些很棒的作品。设计或许很完美，却在无法控制的情况下，未达成预期的结果，甚至连一半都达不到，这可能是因为社会因素、政治因素或其他问题，导致几乎完全无法表达出设计的目的。我想问，在此情况下，当调整不符合原本的期待时，他会完全放弃其设计吗？他会选择放弃，还是会与修改妥协？或是，第三种情况，既能表达设计理念，也会完成作品？您的答案也许能为中国设计师提供一些想法，并协助他们进行未来的工作。在我国，我们有自然风光和景观美化工程，自然风光总是很美，却在进行着景观设计的工作。相较于 80 年代和 90 年代的景观，近年来此领域经历了很大的变化，我真的很感谢景观设计师让我们的城市空间变得更适合生活。我认为景观是最重要的部分，就像美丽的天坛和历史中的一些中国古典园林，我们的祖先在那个年代创造了辉煌的作品，我相信，今天的景观设计师也能创造出伟大的设计。但是过程中却存在着这些问题，我想知道，拉茨先生如何解决这些问题，也许可以稍微启发中国的景观设计师。

拉茨：我很遗憾无法为您提供一个非常不同的经验，即使身为大学教授和系主任，我同样会面临这些问题。如果您无法说服客户，就无法建造任何东西，另一方面，如果做出来的部分不符合计划时，我将会采取法律行动捍卫我的姓名权。

王向荣：现在已经超过 16:00 点了，我们再征求最后两个问题。

发问者 7：我想请问一个简单的问题，拉茨先生的公司规模多大？如何经营公司？

拉茨：官方说法是，我和我太太目前没有自己的公司，从 2011 年起，就由我们的儿子担任公司的负责人。

但是我仍然需负责一些具体的工作。我们公司约有 20 人，其中

In the first instance, this type of park is made for the people who live within a distance of 15 to 20 kilometers, but it has developed into an attraction for tourists.

In Germany, 60 – 90 percent of the people might think of vineyard villas as the ideal landscape, I myself love Italian gardens, but industrial production is a very important part of our culture. We can't hide it and make a green carpet, which covers everything. We have to deal with it, also with the pollution.

Xiaolin Wang: From your comments about this project, the major users are the local people. But as time goes on, maybe the next generation will live there, and the old generation will be gone. What do you expect the next generation will appreciate about your project? Or what is your major expectation for the young generation about this kind of new landscape? What if the history and the process of its creation is not passed on. In a decade will it merely be appreciated as a beautiful landscape. If it is the case, how do you suggest young landscape architects deal with this problem? Or is it a problem?

Latz: It was surprising for me that the young generation was the first one to enter the place. It was really astonishing that 12-17 year old girls and boys were there. For them we created a special space of 15000 square meters in the former manganese ore depot. We transformed 6-7 boxes for sports and play or just sitting and having a chat with other teenagers. The young people came to this place from the other side of the river to meet their friends. Besides numerous small "play-points" we provided two large children's playgrounds which also attracted people from far away. Therefore I am pretty sure that appreciation won't be a problem for the next generation.

As far as young landscape architects are concerned, they get very well trained in our country – even in master courses -- for the dealing with postindustrial landscapes.

Xiaorong Wang: The audience may now pose questions. You can speak either in Chinese or in English.

Q1: Like your project in Israel, This site contained methane in the land. Methane is flammable and can cause explosions. It is also a Greenhouse gas. My question is do you have a method to reuse or dispose of methane in landfill project.

Latz: Of course, using a hermetic system, the biogas gets collected and then transported two kilometers to a plant, where it is used as a heat source.

Q1: It seems that your process is to first treat the waste on the landfill and then to build a new landscape on the existing site. I want to know if you have any suggestions about disposing of the waste and bringing the land back to its original scenery or if there is some technique or method which can help accomplish this.

Latz: As you can imagine it is impossible to move a mountain consisting of household waste and 65 meters high to another place. Besides, wherever you will bring it, it will remain garbage.

Gao: I understand his question, he wants to know if there are alternative methods of disposing of the waste, such as burning?

15人曾受过训练，并有硕士学位的景观建筑师。办公室同仁的自律性很强—组织每一个礼拜聚会一次，即使我的儿子、我或我太太不在，同仁也会以最可靠而有效率的态度工作。

发问者8：在中国，大多数城市的人口密度很高，会在景观工程的施工过程中造成各种问题，对于在现代大都市进行景观工程，您有何建议？

拉茨：抱歉，我也许有些想法，但是我在中国的问题方面没有足够的经验，因此无法提供建议。

王向荣：谢谢。再见。□

Latz: For dumps of this type incineration is rarely used as the burning process needs time and the moisture content of the garbage is high, even in a dry climate. It is in fact much higher than might be expected, so incineration is usually not possible. On the other hand, a new recycling plant is going to be built at the foot of the mountain where a great deal of the garbage will be recycled, and part of that process involves burning technology.

Q2: I am a reporter from China Environmental Newspaper and there are some questions I want to ask. Firstly, is this your first time travelling in China? In this morning's lecture you mentioned that you were highly enlightened by landscape planning and city construction in China. So could you comment more on these topics? Are there any cases you would like to criticize or appraise?

Latz: The Chinese landscape has affected the European landscape in at least two aspects. The first is that for the last 150 to 300 years, China has had a great impact on the theory of European gardens. This is especially obvious in gardens built in the English garden style，which was highly influenced by the dissemination of pictures and prints of Chinese gardens. The second aspect relates to the fact that the diversity of plant species in China is three times as high as in Europe. Introducing Chinese plants had a great impact on the flora used in European gardens and parks.

When you come to China yourself, you are on the one hand impressed about the high quality of landscape and town planning, but on the other hand shocked about the breathtaking rapidity of construction and deconstruction. What you see is quite different from what you have learned from books or the media. The first time I came to China, I lived in a historic two-story hotel which had the look and the atmosphere I knew from literature. This building is now gone. I think the speed of the implementation of projects in Europe is at most one tenth of that in China.

Q3: I want to ask Mr. Latz a question. For every project, you must have your own expectation, but how do you control your projects to achieve your goals. Just like the mine project you mentioned, you want to keep the original look of the mine and to let the site retain its industrial memory. However, in the end, teenagers vandalized the old equipment maybe because they don't relate to this memory. You also said that teenagers are the main users of this park, so how do you deal with the expected users? Do you have some opinion about this problem?

Latz: May I clarify some misunderstandings: I talked about the transformation of a blast furnace site, not of a mine. There was an old mine on the other side of the motorway, but the shaft has been closed and a big tar lake which has nothing to do with the mine has been sealed. I did not speak about this area which is rarely used and left to natural succession.

There was some vandalism but only in the beginning. As just explained, especially the youth appreciates the industrial heritage. Teenagers are not the main users of the park, but the main and only users of their territory.

Translater: Do you mean being certain about who are the park's users?

Q3: What I mean is to what extend can you control the expected results. Just like the children' playground is designed for serving children and the mine project is designed for keeping the city's memory. However, in the end there are teenagers using the park. Would that be a conflict with the original idea?

Latz: The place is designed for everybody. It can be used all the time, 24 hours a day, 365 days a year, it's not just for one group of people.

Xiangrong Wang: You are a designer and so is your wife. Do you often talk about your work with your family and how do you find a balance between work and family? Because when family life is mixed with work life, you end up working all day and that can be annoying. Have any disagreements you and your wife may have had about work entered into your family life?

彼得·拉茨大师报告会签名现场

Latz: This is a very good question, we can never find the most ideal equilibrium. For us professionals it is a very typical characteristic. As a designer you are most of the time thinking about your projects, that is, you can't put the pen down and say I've now finished my work. Even when you're on vacation you cannot truly keep apart your family life and your work. But my wife, my son and I can work together very fruitfully and I think this is good in many ways.

Xiangrong Wang: I also ask this question to professor Wang.

Xiaolin Wang: This question can be simply answered. My husband is a professor as well as an architect. We work in both architecture and landscape and often have disagreements. My answer is very simple, it is very important to face the problem positively. Over the years we have learned to argue from opposite points of view to gradually come to some common idea. Landscape and architecture are often integrated and cannot be treated separately. Landscape is about the environment while architecture addresses a single or a group construction. Which creates solid space. Therefore, landscape and architecture work just like two sides of a mirror. But our landscape designers do not have as strong a concept of space as architects. This is a discrepancy brought about by our training. However, landscape designers need to think on a global scale. So from this point of view, I think from a big to a small scale while my husband thinks from a small to a big scale. Because of this I am good at using logic to deal with big problems while he is more creative. So one of us uses the left brain and the other uses the right and finally we combine together as a unit.

So I would like to suggest to every designer that you should not only learn to analyze and deal with things in an objective and logical way but also learn to cultivate your own creativity because you can't avoid this in the working process. Traditional architecture education always emphasizes the personal style, which leads to different schools. So it may be complementary to come from different majors, but the bad side is we are always arguing. When our children were little they didn't understand why we were always arguing. But when they grew they understood the situation. Now our family travels are like a professional investigation, we see landscapes, we see architectures, we see everything. And we finally become a unit.

Q4: I want to ask a question. In our environment, we find lots of abandoned areas that are seriously polluted. These areas may bring great harm to local residents. So I want to ask your suggestion about eliminating this kind of pollution as well as creating a secure living environment both for local people and wildlife.

Latz: Your question addresses the aspect of these projects which needs to be discussed most. But I don't think that we have a unified, consistent answer, because each site is different. For example, a pharmaceutical factory may have some unique kind of pollution. So you have to do very serious research on every site in order to understand the specific situation. What is definitely clear is that in 90% of the cases the site cannot be developed into residential land.

Q5: But this happens in today's China, these polluted land is

会后彼得·拉茨接受媒体采访

developed into residential areas.

Translator: How can we gain people's trust when these things happen?

Q5: Yes.

Latz: That is really a question, which can be answered only locally and depending on the people's experiences and knowledge.

Xiangrong Wang: The students question are very good and show a concern for human beings. It may be too serious and difficult a question to easily answer. Let me change the question and relax for a moment. Now I have a principle, that is do not talk about work when you're at home. Because we will argue with each other without coming to an agreement. It is fine in the office because it is normal to have arguments in the office.

The question I asked him is that his home is also his office, so it maybe more complicated and I also have no good idea. Many of my friends and colleague have families like these so I can gain some experience from them.

Xiaolin Wang: It needs both insistence and forgiveness, like Professor Wang said. Sometimes if you only know insistence then you won't handle the problem.

Yang: I'm now translating a book of Mr. Latz and I also read some chapters and learned a lot from it. I think landscape architecture can really turn the foul and rotten into the rare and ethereal. But on the other hand, it can actually turn magic to decay, which is the case with some beautiful scenic spots and natural heritage sites. So I think the key point is how we act, how we plan and design to express the charm of landscape architecture. You can see from his lecture this morning as well as in some of his books that through his design, a cold industrial building can turn into a warm place, It seems like these works have their own emotions. A place with cold and dead metal can turn into a lively and vital place with plant design and so on. We can see many cases these days which includes such spirit. I actually think this is exactly what landscape architecture is, or in other words, the charm of this science.

My question is how he brings this kind of spirit and emotional factors into his planning design of industry land. I have had some contact with him. I feel he is a serious German from his appearance. But in some of his works I can feel a kind of effusion of his inner emotions, the emotions come from his heart. That is the question I want to ask.

Latz: Of course I want to avoid having no emotion. But I think it is the most important task to find and develop the structure which is crucial for a project. Otherwise it would be impossible to keep the control over all the objects you or somebody else includes. Especially if you work with large dimensions, a strong structure will enable to decide on the choice of the elements and the rules of their arrangement. Objects with different appearances can be well integrated into this rather neutral structure.

The use of the elements can be decided by the future users and visitors. We also say what might sound like a joke: "The photographer decides whether they are beautiful or not".

But I must mention one point explicitly, that is the maintenance. It is very important for the whole development process. For example, when you prune a tree it is against its natural structure. Its semantic quality or expression will change the space too.

Maintenance work means to control what may not be changed by others. It ensures what kind of state and space you want people to see in later years. Without maintenance, the place can loose a lot of information and elements.

Landscape architects never produce plans showing how a site will look after one year. Usually we describe the state after 15 years or even a longer period that means we draw mature trees. You cannot avoid natural disasters or other factors that change your design, so it is not possible to try to control everything completely.

Q6: I have worked in the Ministry of Construction for over 30 years and have always been engaged in this field. I have contacted a lot of designers who have confusion about their work. I think it may be a problem relating to the country's system. Our country's system is totally different from other countries'. Our designers often produce brilliant work which is designed on the basis of the local situation, local culture, history, environment and social background. Maybe it is a perfect design but the project can not be controlled enough to achieve the expected outcome, even for 50 percent of the design. It may due to social reasons, or political reasons or some other problems, so the goals of a design are almost never fully expressed. Under such a situation, I want to ask, does he totally give up on his design when a modification is not like what he expected? He Would he choose to give it up, or will he compromise on the modification, or in the third case, can he both express his design idea and complete the project as well? Maybe it can help our Chinese designers' concepts and future works. We have natural scenery and landscape works in our country, the natural sceneries is always beautiful. But the landscape design has been a work in progress. Comparing the landscapes from the 1980s and 1990s, the field has undergone great changes recently. I really appreciate how our landscape designers can make our city spaces becomes better for living. I think landscape is one of the most important parts. Just like the beautiful Tiantan and some Chinese classical gardens in history, our ancestor created such brilliant work in that age. I believe today's landscape designers could also create great designs. However there are these problems that exists in this process. I want to know how Mr. Latz deals with these problems and possibly inspire Chinese landscape designers somewhat?

Latz: Unfortunately, I cannot describe to you a very different experience. Even as a professor and chair at the university, I have always had these problems. If you cannot convince the client you can build nothing. On the other hand, if the realized parts of the project don't conform to the plans, I can take legal action against the use of my name.

Xiangrong Wang: Now it is past four o'clock. Let's have two last questions from the audience.

Q7: I want to ask a simple question, what is the scale of Mr. latz' company and how does he run his company?

Latz: Officially, my wife and I have now no company of our own. Since 2011, our son has been the head of the office.

But I am still responsible for some specific projects. In our company there work around 20 people, 15 of them are trained landscape architects with a master's degree. The office has a strong self – organization having a weekly get-together every Monday. Even if my son and/or my wife and I are not present, the staff is working most reliably and effectively.

Q8: Most cities in China have a high population density and it causes various problems in the construction process of landscape projects. Do you have any advice concerning landscape projects in the modern metropolis?

Latz: I am sorry to say that I would perhaps have some ideas, but I have not enough experience in China for handing out advice.

Xiangrong Wang: Thanks. See you.■

大师报告会对话嘉宾简介：
Dialogues with the guests at the presentation:

嘉宾主持：王向荣 教授／北京林业大学园林学院副院长

Guest: Xiangrong Wang Professor / Assistant Dean of Landscape Architecture School, Beijing Forestry University

嘉 宾：王小璘 教授／台湾造园景观学会名誉理事长／世界园林总编

Guest: Xiaolin Wang Professor / Honour Principal of Chinese Landscape Architect Association in Taiwan / Editor–in–Chief of *Worldscape*

嘉 宾：高 翅 教授／华中农业大学副校长

Guest: Chi Gao Professor / Vice President of Huazhong Agricultural University

嘉 宾：杨 锐 教授／清华大学的景观建筑系系主任

Guest: Rui Yang Professor / Head of Landscape Architecture Department, Tsinghua University

翻 译：郑晓笛 博士／美国注册风景园林师（宾夕法尼亚州）／清华大学建筑学院博士在读／中国花卉园艺与园林绿化行业协会国际部主任

Translator: Xiaodi Zheng Ph.D / Registered Landscape Architect in PA, USA / Doctoral Candidate at School of Architecture in Tsinghua University / Director of International Affairs,Chinese Flowers Gardening and Landscaping Industry Association

刘纯青、汤名洁（中译）、张勤、骆思同（编译），彼得·拉茨（校订）

Translated and edited by Chunqing Liu / Jessie Tang / Qin Zhang / Sitong Luo, English reviewed by Peter Latz

宣示权利

以色列特拉维夫市夏隆公园海瑞亚垃圾填埋场再利用

LAYING CLAIM

HIRIYA LANDFILL RECLAMATION ARIEL SHARON PARK, TEL AVIV, ISRAEL

尼尔·柯克伍德　Niall Kirkwood

图 01 越过海瑞亚垃圾填埋场看特拉维夫的天际线
Fig 01 View towards Tel Aviv Skyline over Hiriya Landfill

图 02 从远处可见海瑞亚垃圾填埋场形成了一个顶部平坦广阔，四周是陡坡的人造地形
Fig 02 Hiriya Landfill forms a man-made topography with a flat expansive top and steeply sloped sides, visible from afar.

图 03 没有很好的覆盖及处理措施，海瑞亚垃圾填埋场底部的垃圾和渗滤液暴露在外
Fig 03 Garbage and leachate at the base of the Hiriya Landfill are exposed without proper capping and treatment measures.

项目概述

夏隆公园是基于一块大型的被干扰和被废弃的土地而再生形成的，它位于以色列最大的城市中心区。该景观设计方案由一个多学科团队提出。这一覆盖了 800hm^2（1975 英亩）的地区的重建，为周围都市区——丹地区的全体居民创造了一个集被动型康乐、生态储备和休闲为一体的区域。在这个新的天然公园核心，坐落着海瑞亚——一座大型的市政垃圾填埋山，它主导了这片平坦的泛滥平原的景观（图 01）。从 1952 年至 1998 年期间，这个垃圾填埋山包含了 1600 万 m^3 的城市生活垃圾。它占地 40hm^2，足有 60 米高，形成了一个有着广阔的、平坦的顶部和陡峭的边缘的人造地形，很远的地方就可看见（图 02）。它是一个符号，既展示着过去对这片土地的滥用，也展示了未来的转型和新的再利用公园的康复。

到 2020 年，特拉维夫都市区域的人口将增长至 330 万，其中

Park Ariel Sharon is based on the regeneration of a large disturbed and currently abandoned tract of land in the midst of Israel's largest urban district. The landscape design proposal by a multi-disciplinary team arising from the regeneration of this site covers 800 hectares (1975 acres) creating a passive recreation, ecological reserve and leisure area for the population of the surrounding metropolitan area, the Dan region. In the core of this new landscape park sits Hiriya, the large municipal landfill mountain that dominates the flat floodplain landscape(Fig.01). Operating from 1952- 1998 and containing 16 million cubic meters of municipal household waste and with 40 hectares footprint and 60 meters in height with a flat expansive top and steeply sloped sides, the landfill forms a man-made topography, visible from

图 04 环绕海瑞亚垃圾填埋场的大片农田和阿亚隆河
Fig.04 Large area of farmland surrounding Hiriya Landfill and route of the Ayalon Stream

图 05 海瑞亚垃圾填埋场的顶部为俯瞰特拉维夫地区提供了一个独特的制高点
Fig 05 The top of Hiriya Landfill offers a unique highpoint to overlook the Tel Aviv area

图 06 海瑞亚垃圾填埋场的核心部位展现了阶梯式填埋及覆盖方式
Fig 06 Inner core of Hiriya Landfill showing benching and capping activities

120万将居住在特拉维夫地区，大约一半的居民将环绕居住在曾经的泛滥平原和垃圾填埋场区域附近（图03）。环绕该区域的包括一些最贫困的地区，到目前为止没有明确的绿地空间，尤其是和特拉维夫北部的亚肯公园相比时（这一缺陷）更加明显。

特拉维夫大都市是一个活泼的、忙碌的、密集的、动态的城市中心。公园将成为一个居民可以在此休闲一个小时或一天的地方，一个可供他们将享受一个安宁、开放的空间，一个有着小树林，原野和湖泊的文化景观。公园还将纳入丰富的文化活动，以丰富城市生活。

该地区的东部和南部，有一条南北向的交通要道和一条东西向的交通要道形成场地的边界，两条交通线在这一地区的东南角相汇。也就是说，从汽车交通运行的角度来讲，这里是以色列的中心点。由于独特的历史环境，这一大型、开放的土地被遗留在了特拉维夫地区的中心地带，具有成为世界最大城市公园之一的潜力，也具有成为一个当地露天型社会中心和会议中心的潜力。

现在，被污染的阿亚隆河正在被修复治理中，成为公园空间的骨干。它穿过湿地、一个湖泊和一个为公园提供水源的水库，并作为居民区的洪水缓冲区。海瑞亚的城市垃圾填埋山将作为一个戏剧性的地形元素，与附近的农田（图04）和考古遗址共同融入到公园中。夏隆公园、悠久的MikvehYisrael农业学校与现有相邻的公园——Darom公园、拉马丹公园，将形成一个城市景观空间体系。

场地分析

这一规划主要包括两块土地：一块约300hm^2（740英亩），位于耶路撒冷——特拉维夫高速公路（1号国道）南侧，包含以色列第一所农业学校——MikvehYisrael的建筑物和土地，该校址是受到法律

afar(Fig.02) and a symbol of both the past abuse of the site and the future transformation and healing in the new reclamation park.

By 2020, the population of the Tel Aviv metropolitan area will have grown to 3.3 million, of whom 1.2 million will reside in the district of Tel Aviv, about half of which will surround the former floodplain and waste landfill site(Fig.03). The urban area surrounding the site includes some of the most underprivileged parts of the district, which today have no significant open space, particularly when compared to North Tel Aviv's Yarkon Park.

The metropolis of Tel-Aviv, is a lively, busy, dense and dynamic urban center. The park will be a place where residents can retreat for an hour or for a day, where they will enjoy the experience of a restful, open space, and the cultural landscape of groves, fields and lakes. The park will also incorporate cultural and enriching activities, complementing city life.

Eastwards and southwards, the site is bounded by the north-south and east-west traffic arteries, crossing at the south-east corner of the site. This could be described, in terms of motor traffic circulation, as the central point of Israel. Due to a unique set of historical circumstances, this large, open tract of land has been left in the heart of the Tel Aviv district, with the potential of becoming one of the world's great parks, as well as a local open-air social center and meeting place.

The existing polluted Ayalon Stream currently under

图 07 垃圾填埋场斜坡的阶梯式处理与建设工程
Fig 07 Benching and construction of landfill slopes

图 08 海瑞亚垃圾填埋场底部的生物反应器和回收中心，该中心提供能量产品和有机和非有机废物的再分配。
Fig 08 Bioreactor and Recycling Center at base of Hiriya Landfill. This center provides energy production and holistic redistribution of organic and non-organic waste

保护的。由于它已经不再作为一所常规的农业学校运营，可以使它的设施和项目现代化，预计它将在公众访问，游路，公共活动方面与前述的大型公共区域合作发展。另一个地区，约 500hm^2（1235 英亩），位于国道的北侧，包括海瑞亚垃圾填埋山、回收中心、Darom 公园、一些农田，这些农田被不同的持有者使用，用不同的方式从以色列土地管理局租赁。该地区还包括阿亚隆季节性河流以及它的冲积平原和一些古迹。

海瑞亚垃圾填埋场顶部（图 05）视野广阔，可以越过特拉维夫市看到波光粼粼的地中海。从另一种意义上说，海瑞亚的视野和拟建的阿亚隆公园对以色列及其后面的国际地形来说都是一个新的典范。

类似海瑞亚的垃圾填埋场和其他形式的工业产品的生产使命已走到尽头，就像许多其它棕地的情况一样，通常被废弃了。怎样使海瑞亚这一个封闭的废弃的垃圾填埋山、一个包围了特拉维夫市的边缘区域 1975 英亩土地、一个长期被遗弃和污染的标志、一个 50 万居民每天经过的地区，重生为一个具有世界最高品质的、充满活力公园的一部分？重生为特拉维夫市及其更远的居民服务的、一个舒适的、象征再生和拯救的符号？与海瑞亚和未来的夏隆公园相关的工作，不只是一个简单的回归到对资源耗尽与贬值的土地的生产性应用——对过去工业环境的整理，而是意义深远的转变。在这转变的过程中，城市、社区、专业人士要对这个有争议的区域形成主张。源自包括拉丁美洲、欧洲、美国的规划师、景观建筑师、生态学家、工程师、艺术家、科学家、建筑师团体与当地的景观设计师形成的景观理念，应对了这个

remediation forms the spine of the park, running through wetlands and a lake, a water reservoir feeding the park, and acting as a buffer against the flooding of residential areas. The Hiriya municipal landfill mound will be integrated into the park as a dramatic topographical element next to agricultural fields(Fig.04) and the archaeological site. Park Sharon together with the historical agricultural school of MikvehYisrael and the existing adjacent parks, Park Darom and the Ramat Gan Park, will form a sequence of urban landscape spaces.

The plan includes two main areas of land: one area, approx. 300 hectares (740 acres) lies south of the Jerusalem - Tel Aviv highway (National Road no. 1) and comprises the buildings and lands of Israel's first agricultural school, MikvehYisrael. As such, it is protected by law, but since it has ceased to function as a regular agricultural school, it may modernize its facilities and programs and it is expected to be developed in collaboration with the larger public area of the site, in terms of public access, paths, public and activities. The second area, approx. 500 hectares (1235 acres) lies north of that road and comprises the Hiriya landfill and recycling center, Park Darom, cultivated agricultural land held by a number of stakeholders under different leases from the Israel Lands Authority, the seasonal Ayalon Stream and its floodplain

图 09 海瑞亚垃圾填埋场再利用设计工作组，特拉维夫。
Fig 09 Hiriya Landfill Reuse Design Charette, Tel Aviv

复杂的问题，并且发展了未来设计的建议，成为与夏隆公园项目相关的新规划措施的一部分。

扩展的建议是将公园的景观基础设施、道路、交叉路口、通向公园的桥梁和农业草地变成一个基础设施投资，而不是简单的工程或园艺技术。规划也开始转向解决这一地区的溪流和土地的环境污染，并且制定了策略，以使周围的居民区、外围社区和偏远的城镇都能享受场地的自然资源。

该公园的发展的基础和原则是基于法定规划（区域纲要规划 5/3）、海瑞亚展览会、2001 年规划研讨会形成的。设计和规划工作是前述工作的一个直接的续集，旨在检查、集成和详细阐述这些理念，将它们转换为实际的规划条款，并为该公园的应用总体规划提供具体内容。这些内容将形成阿亚隆公园总体规划的基础，那将使公园进一步发展，并为其它部分的规划进行准备，这是基于分步实施的考虑。

《博物馆中的海瑞亚：艺术家和建筑师为地区康复的提议》展览，在特拉维夫艺术博物馆举办，从 1999 年 11 月持续至 2000 年 4 月，由 Beracha 基金会发起。展览的目的是创造一个机会来构思一个建议：如何将曾经的海瑞亚废物堆变成一个独特的场所。这一展览对提高公众意识做出了贡献，并将讨论扩展到规划和环境之外的方面。不同的建议包括一些敏锐的社会评论和一些原始甚至是乌托邦的设想。该展览强调了海瑞亚废料堆从一个严重的环境问题，转变成夏隆公园规划的一个关键组成部分的可能性。

2001 年 1 月举行的一个国际性的工作研讨会“向着阿亚隆”，重

and Archeological sites.

The view from the top(Fig.05) of the Hiriya landfill is expansive reaching beyond the city of Tel Aviv to the sparkling Mediterranean Sea. In another sense, the view from Hiriya and the proposed Park Ayalon is a new paradigm both for Israel and the international terrain beyond.

The productive life of landfills such as Hiriya and other forms of industrial production come to an end, and are usually replaced by abandonment and dereliction as is the case of so many brownfield sites. How does Hiriya the closed waste landfill and surrounding 1,975 acre lands on the edge of Tel Aviv, long a symbol of dereliction and pollution to the half million citizens who pass it every day, be reborn as part of a vibrant park of the highest quality in the world and an amenity and symbol of rebirth and healing for all the residents of the city and beyond? The work associated with Hiriya and the future Park Sharon is not just a simple return to the productive use of exhausted and currently undervalued land- a tidying up of the past industrial environment, rather it signals a profound shift in the way in which cities, communities and professionals must lay claim to this disputed area. Speculative yet practical landscape ideas from an international group of planners, landscape architects, ecologists, engineers, artists scientists and architects from around the world- Latin America, Europe, United States- collaborated with local landscape practitioners in Tel Aviv to address this complex question, and have advanced future design proposals as part of new planning initiatives related to the Park Sharon project.

Broader proposals where landscape infrastructure, roadways, intersections, bridges to the park and agricultural meadows become an infrastructural investment rather than simply engineering or horticultural techniques. Planning has also moved on to tackle the environmental pollution to the site's streams and soils and outlined strategies to bring surrounding neighborhoods and outlying communities and remote towns to enjoy the natural resources of the site.

The statutory plan (Regional Outline Plan 5/3), the Hiriya Exhibition, and the 2001 planning workshop laid down foundations and principles for the development of the park. The design and planning work was a direct sequel, intended to

图 10-11 海瑞亚垃圾填埋场设计工作组，包括景观设计师尼尔·柯克伍德和艺术家米勒·乌克勒。
Fig 10-11 Hiriya Landfill Design Charette, Members of Charette on landfill including landscape architects-Niall Kirkwood and artist Mierle Laderman Ukeles.

图 12 工作组景观设计师彼得·拉茨所绘草图
Fig 12 Hiriya Design Charette Sketch by landscape architect Peter Latz

在强调坐落在国家中心的这个尺寸不同寻常的公园，以及它涵盖了娱乐、休闲、运动及教育于一体的多功能场所的角色。夏隆公园将会提供一个创建都市级别的娱乐空间的机会，像其他著名的公园一样，成为城市的一种符号。工作组的指导方针公布于 2001 年 11 月，包括建议未来所要采取何种行动，例如：建立公园管理和组织机构，确定公园管理的管理者，资金筹措和启动的进程等。它将会被设计成为一个服务内容更广泛的公园：最终，将被视作空间连接体，将海洋与大都市、本－古力安国际机场以及耶路撒冷连接统一为一个整体。公园将象征一个文化意识方面的巨大转变，表达一种责任，致力于改变和治愈我们这个受到太多污染的当代世界（水源，空气，噪音，固体垃圾等等）。公园将具有生态意义和教育意义。在某种程度上讲，这个公园是一个巨大的城市实验场。

项目开发

公园的规划必须被看做是一个持续不断进化的项目，而不只是一个单一的概念（图 06–07）。

海瑞亚垃圾填埋场是公园设计中一个主要的元素。海瑞亚应被视为公园中一个主要的、象征性力量，为了在这里全面逆转那么巨大的退化，创造一种朝向变化与治愈的文化模式的转变。公园的计划包括回收中心 (0)（图 08），它将确保海瑞亚最终完整地完成恢复健康土地的目的。

公园的土地不属于任何市政当局，它受到国家和地区委员会管辖。1998 年，区部委员会的内政部形成一个法定规划——区域纲要计划 5 号，目的是为了维护这一至关重要的土地的绿色未来，保护这块土地，直到详细公园计划付诸应用和实现。

法定规划是结构化的，作为一个总体框架，而不是作为一个特定的区域规划。基于这一法定规划的原则是所有公园范围内的农业土地临时再定义为“供公园使用”以作储备之用。这一措施由于法定所有权的复杂性已经被接受。特定的用途将在随后的阶段决定。另一个基本原则是逐步实施计划，在土地租赁合同到期之前允许每个承租人继

examine, integrate and elaborate these ideas, to translate them into practical planning terms, and to offer concrete contents towards the formulation of an overall applied master plan for the park. These contents will form the basis of the Park Ayalon Master Plan, which will enable further development of the park and preparation of plans for its various parts, taking into consideration that execution will be in phases.

'Hiriya in the Museum: Artist's and Architect's Proposals for Rehabilitation of the Site', held at the Tel Aviv Museum of Art, November 1999 – April 2000, was sponsored by the Beracha Foundation. The purpose of the exhibition was to create an opportunity to conceive speculative proposals on how to transform the former Hiriya waste dump into a unique site. The exhibition contributed to public awareness and extended the discussion to other disciplines, besides planning and environmental aspects. The different proposals included some astute social critiques and original, even utopian, suggestions. The exhibition highlighted the potential for the transformation of the Hiriya waste dump from a severe environmental problem to a key component of the Park Sharon plan.

"Towards Ayalon", an international workshop held in January 2001, emphasized the unusual size of the park, its location in the heart of the country, and its diverse roles as a place for recreation, relaxation, sports and education. The Park Sharon will provide an opportunity to create a recreational space at a metropolitan level, like other renowned parks that have become symbols of their cities. The workshop's guidelines were published in November 2001 and included recommended actions to be taken in the near future such as: establishing park management

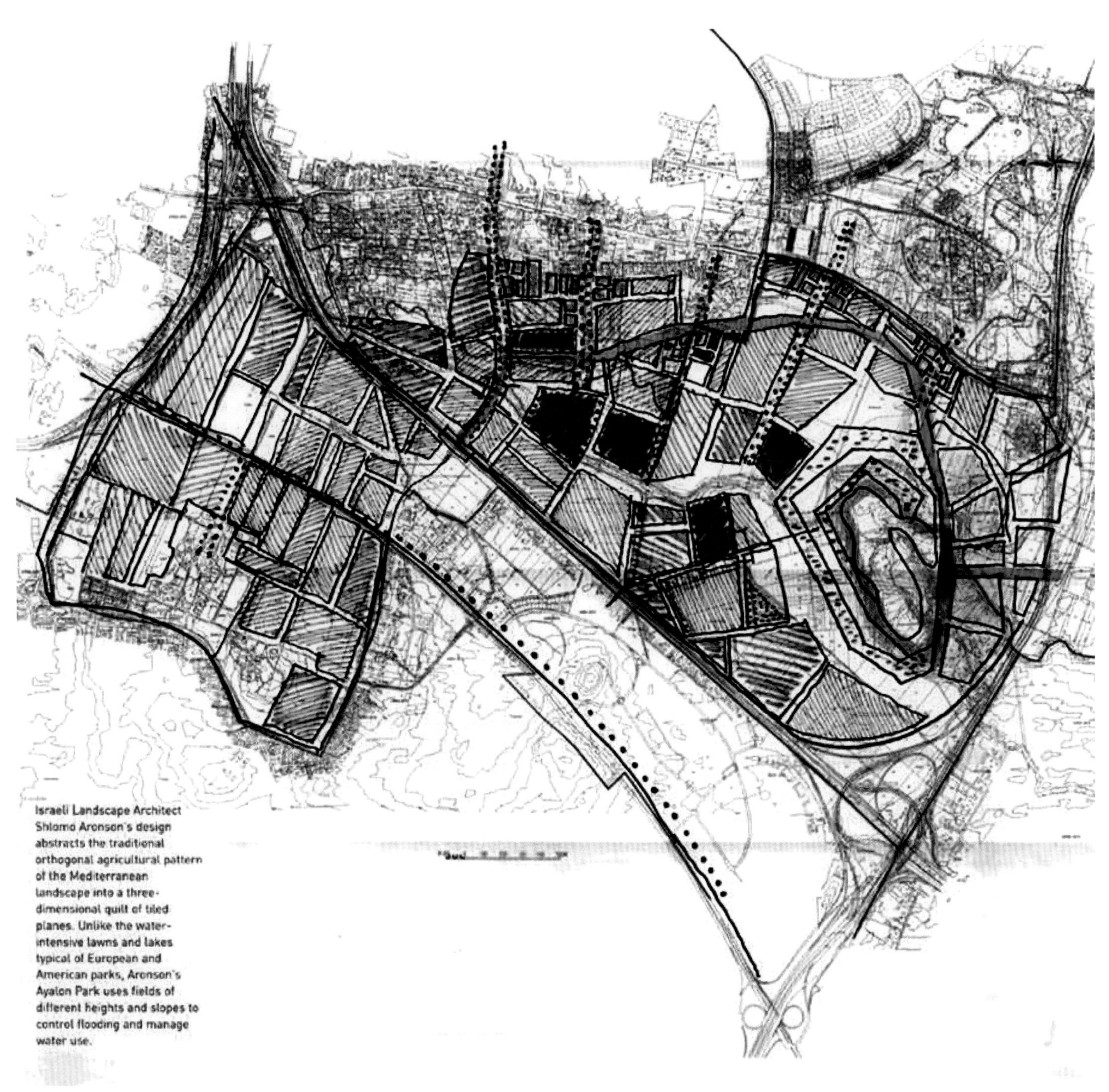

Israeli Landscape Architect
Shlomo Aronson's design
abstracts the traditional
orthogonal agricultural pattern
of the Mediterranean
landscape into a three-
dimensional quilt of tiled
planes. Unlike the water-
intensive lawns and lakes
typical of European and
American parks, Aronson's
Ayalon Park uses fields of
different heights and slopes to
control flooding and manage
water use.

图 13 工作组景观设计师什洛莫·阿伦森绘制的草图
Fig 13 Charette Sketch by landscape architect Shlomo Aronson

续土地种植。因此，公园内的几大分区在未来几年内仍作为耕种地使用。

参与了早期的设计工作的风景园林师、设计师和艺术家包括墨西哥的马里奥·谢赫楠，德国的彼得·拉茨，美国的朱莉·巴格曼，米尔勒·拉德曼·尤克外斯，艾丽莎·罗森伯格和肯·史密斯，以色列的什洛莫·阿伦森，普莱斯奈尔，雅艾尔·摩瑞亚，玛雅·沙菲尔，施洛米泽维和大卫·古根海姆，还有生态学家罗伯特·法朗士（哈佛大学）和史蒂文·汉德尔（罗格斯大学）也参于了本次设计。整个规划设计工作由作者和纽约的风景园林师劳拉·斯塔尔组织，马丁·韦尔博士负责的贝拉恰基金会和特拉维夫规划部门办公室（TVDPO）的纳米奥·安吉拉实施监督，同时受到了丹尼·斯特恩伯格和祖瑞特欧荣的帮助（图 09–16）。海瑞亚垃圾填埋山再利用设计竞赛由来自德国的彼得拉茨和他的团队在 2004 年 9 月获得中标（图 17）。作者是特拉维夫市竞赛评审团主席。目前，海瑞亚再利用的项目仍在进行中。□

and organization, the park administrator, funding and launching the process. The park will be designed with the aim of serving a context broader than itself: ultimately, it will be considered as part of the spatial continuum between the sea, the Metropolis, Ben-Gurion International Airport and Jerusalem. The park will symbolize a dramatic shift in cultural consciousness, expressed as a commitment to transform and to heal various polluted and degraded aspects of our contemporary world (water, air, noise, solid waste, etc.) The park will be ecological and educational. In a sense, the park is a large-scale, urban laboratory.

program development. The planning of the park must be seen as a continuous and evolving process, rather than a monolithic concept. (Fig.06-07)

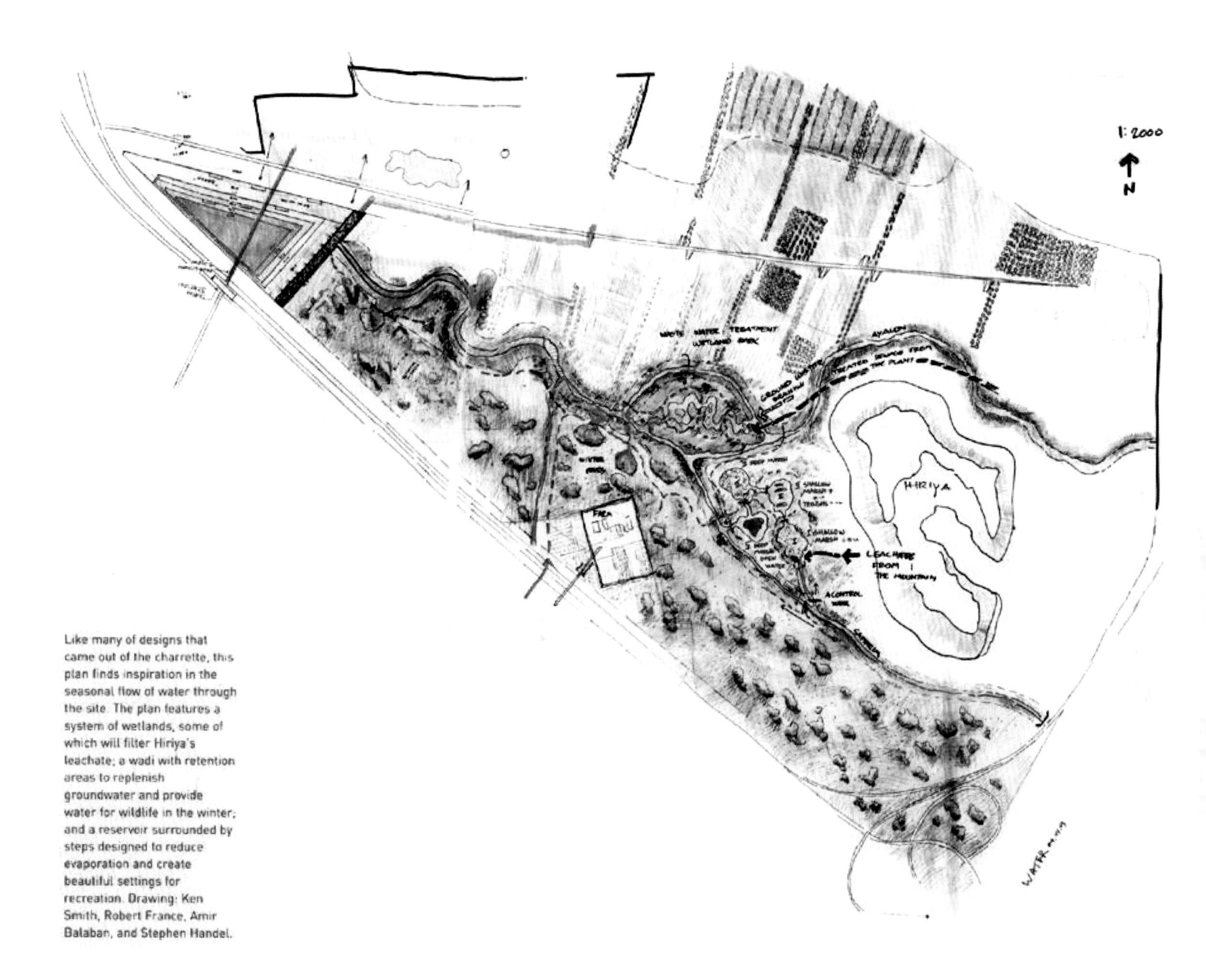

1:2000
N
HIRIYA
Like many of designs that came out of the charrette, this plan finds inspiration in the seasonal flow of water through the site. The plan features a system of wetlands, some of which will filter Hiriya's leachate; a wadi with retention areas to replenish groundwater and provide water for wildlife in the winter; and a reservoir surrounded by steps designed to reduce evaporation and create beautiful settings for recreation. Drawing: Ken Smith, Robert France, Amir Balaban, and Stephen Handel.

图 14 工作组景观设计师马里奥·谢赫楠绘制的草图
Fig 14 Hiriya Design Charette Sketch by landscape architect Mario Schjetnan

Hiriya landfill is a major element in the park design. Hiriya should be seen as a major, symbolic force in the park, creating a cultural paradigm shift towards transformation and healing, in order to totally reverse the great degradation that has occurred here. The park's program including the recycling center (0) (Fig.08)will guarantee commitment to Hiriya's eventual, complete restoration as healthy parkland.

The park land does not belong to any municipality; it is under the jurisdiction of the State and the District Commission. In 1998, the District Commission of the Ministry of the Interior initiated a statutory plan, Regional Outline Plan No. 5/, in order to secure the green future of this vital piece of land, and safeguard it until detailed park plans can be applied and realized.

The statutory plan is structured as a general framework and not as a specific zoning plan. The principle upon which the statutory plan is based on is a temporary redefinition of all the agricultural lands within the park's boundaries as reserve for 'park use'. This measure has been taken due to legal ownership complexities. The specific uses will be determined at a later stage. Another basic principle is the gradual implementation of the plan, allowing each lessee to continue cultivating land until the lease term has expired. As an outcome, several large plots within the park may remain cultivated for years to come

Landscape Architects, designers and artists involved with the early design work included Mario Schjetnan from Mexico, Peter Latz from Germany, Julie Bargmann, MierleLadermanUkeles,

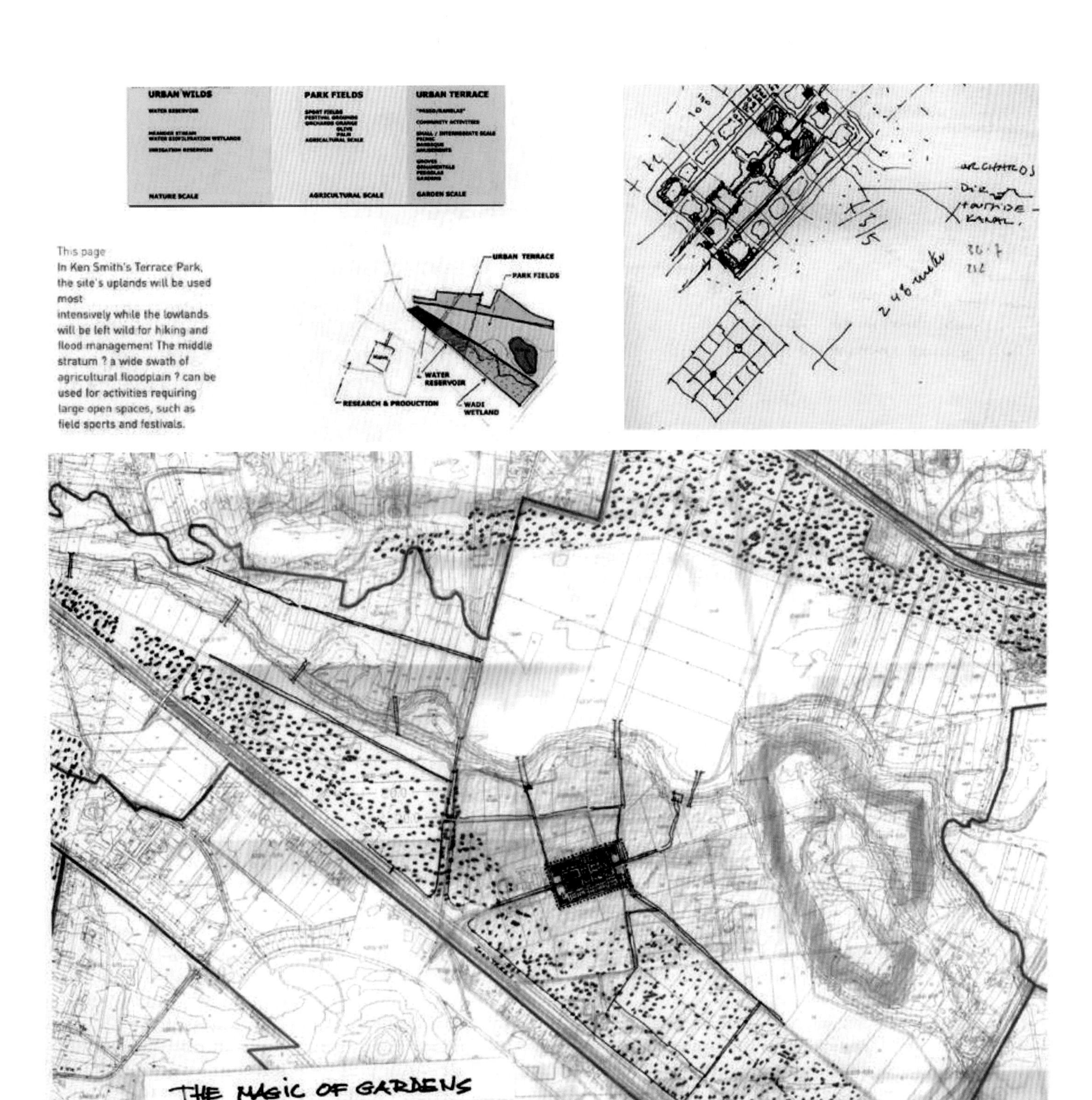

This page
In Ken Smith's Terrace Park, the site's uplands will be used most intensively while the lowlands will be left wild for hiking and flood management The middle stratum ? a wide swath of agricultural floodplain ? can be used for activities requiring large open spaces, such as field sports and festivals.

图 15 工作组景观设计师彼得·拉茨绘制的草图
Fig 15 Hiriya Design Charette Sketch by landscape architect Peter Latz

ElissaRosenburgh and Ken Smith from the United States, Shlomo Aronson, UlrikPlesner, Yael Moria, Maya Shafir, ShlomiZeevi, and David Guggenheim from Israel with ecologists Robert France (Harvard) and Steven Handel (Rutgers) also participated. The planning and design work was organized by the author and landscape architect Laura Starr of New York and overseen by Dr Martin Weyl of the Beracha Foundation and Naomi Angel of Tel Aviv Department of Planning Office (TVDPO) with assistance from Danny Sternberg and ZuritOron. (Fig.09-16)The competition to design the Hiriya Landfill Reclamation was awarded to Peter Latz & Partner of Germany in September 2004(Fig.17). The author was chair of the competition jury based in Tel Aviv. The Hiriya Reclamation project is still ongoing.■

图 17 海瑞亚设计竞赛评委以及马丁·韦尔博士和尼尔·柯克伍德教授的合影
Fig 17 Hiriya Design Competition Jury including Dr Martin Weyl and Professor Niall Kirkwood

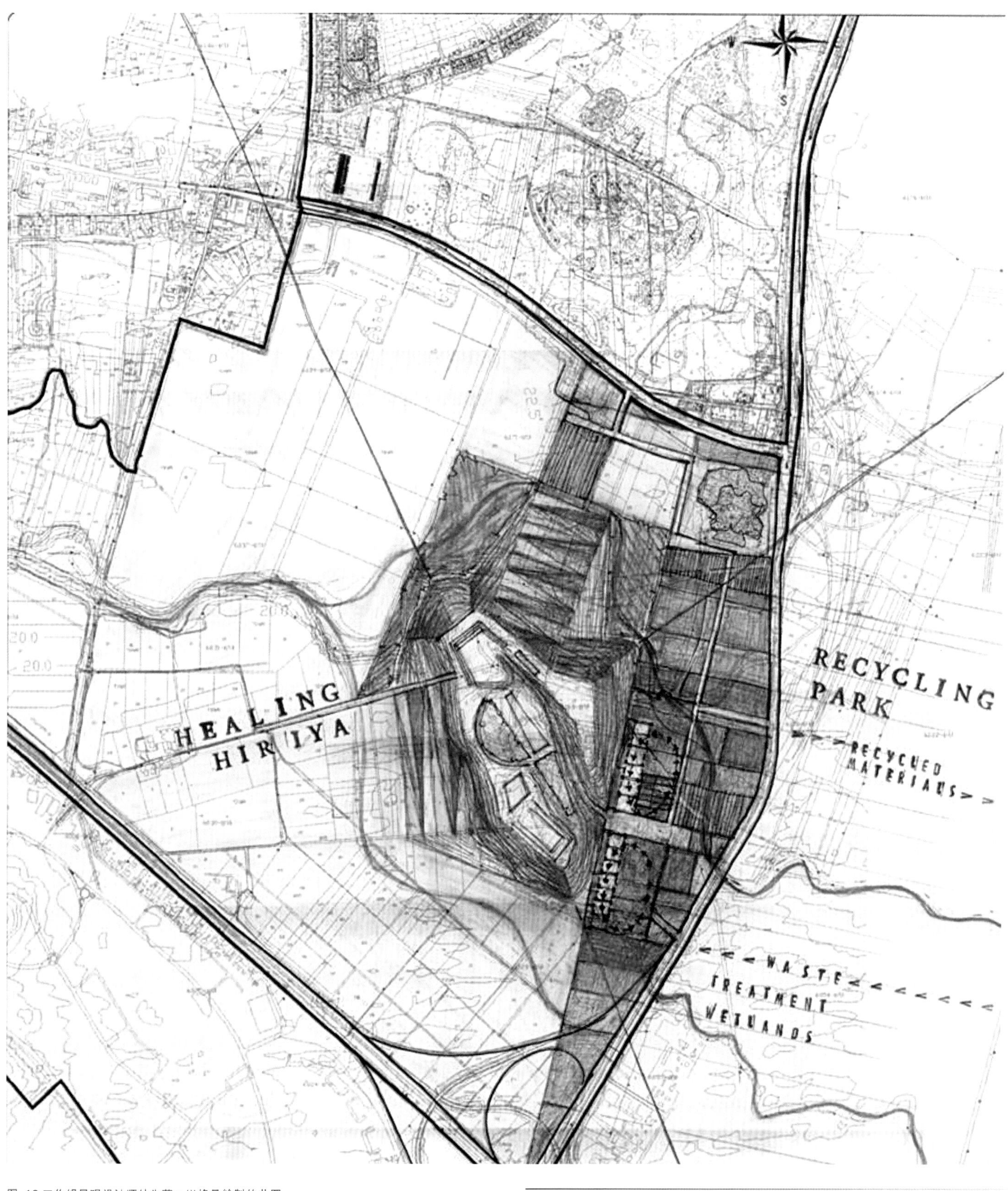

图 16 工作组景观设计师什朱莉・巴格曼绘制的草图
Fig 16 Charette Sketch by landscape architect Julie Bargmann

作者简介：

尼尔・柯克伍德 / 男 / 博士 / 美国景观设计师协会会士 / 哈佛大学景观设计与技术教授

Biography:

Niall Kirkwood/male/DSc/FASLA/Professor of Landscape Architecture and Technology,Harvard University

张红卫（中译）李敏（校译）
Translated by Hongwei zhang, Reviewed by Min Li

什么使美国纽约清溪垃圾填埋场的改造备受瞩目？

WHY DOES FRESHKILLS PARK ATTRACT SO MUCH ATTENTION?

郑晓笛

Xiaodi Zheng

图 01 废弃的垃圾填埋场挖掘机，计划以后被用于支撑清溪公园的入口标志。
Fig 01 Old landfill machinery which will be used to hold signage boards as a landmark in the future.

1. 引子

近十年来，在全球风景园林专业内引起最普遍关注的垃圾填埋场再利用项目无疑是美国纽约州清溪垃圾填埋场改造为公园的项目，其被称为纽约市近百年来建设的最大的城市公园项目，占地面积约为纽约中央公园的3倍大。2001 年3 月22 日，最后一驳船的垃圾沿着清溪运抵了清溪垃圾填埋场并载入了史册，成为该填埋场封场的标志性事件。2012 年4 月，笔者专程走访了清溪垃圾填埋场。在宣布封场11 年后，该场地仍然未对公众全部开放，只能通过专门预约或者参加定期组织的由专人带领的场地游览项目进入填埋场内部（图01–02）。目前进入到施工阶段的景观项目，只是场地外围一处占地不足1 hm^2 的社区公园(图03–06)和一处占地约8.5 hm^2 的运动场地。那么，到底是什么使这个项目吸引了全世界的目光呢?

2. 热议的“话题”

首先，清溪垃圾填埋场本身的规模和角色引来了全世界的关注。凭借890hm^2 的占地面积，它曾经是世界上最大的卫生垃圾填埋场（图07）。在其宣布封场不到半年的时间，纽约市政府于2001 年9 月5 日宣布举办国际竞赛征集各国设计师们对于将其改造为清溪公园的构想。然而仅仅在6 天后，震惊全球的911 事件爆发了。在灾后清理的过程中，清溪垃圾填埋场的西区被重新开启以消纳部分911 事件所产

1. Introduction

Over the past decade, Freshkills Park in New York City has undoubtedly been the most discussed landfill reclamation project within the profession of landscape architecture. It will be the largest public park built in New York City over the last century with an area almost three times the size of Central Park. On March 22nd, 2001, the last barge of garbage was shipped to Fresh Kills Landfill marking its official closure date. In April 2012, I paid a visit to the landfill. Eleven years after the announcement of the closing of the landfill, the site has yet to be opened to the general public.

图 02 进入清溪垃圾填埋场设有安保的入口，可以看到上方“禁止进入”的标识牌。
Fig 02 Approaching Fresh Kills Landfill through controlled security point.

图03–06 施穆公园位于北区垃圾堆体的北边界，占地不足1 公顷。它是一个社区公园，设有一个颜色丰富的儿童游戏场地、几个运动场和一个服务中心。在笔者参观这个公园的时候（2012 年4 月），公园的施工已经接近尾声，即将对公众开放。
Fig03–06 Schmul Park is located on the northen edge of the North Mound, which occupies an area less than 1 hectare. It is a community park composed of a colorful playground, several sport courts, and a comfort station. At the time the author visited the park (April 2012), its construction was about to complete and will be open to the public very soon.

生的废墟垃圾。这一事件使清溪垃圾填埋场对于纽约人而言被赋予了更深的意味。它不再仅仅是堆放厨余垃圾、旧家具、破衣服等生活垃圾的一个令人厌弃的场所，而承载了震惊、悲伤、愤怒等特殊的情感。部分911事件的废墟被运至清溪填埋场进行分拣和处理，全世界的目光曾经聚焦在这里。至今，还可以在场地上看到当时为此搭建的白色临时构筑物（图08），令人不禁联想到这里曾经见证了怎样的历史？这些废墟诉说了怎样的故事？这里掩埋了怎样的情感？这个填埋场的未来如何？它应该被开发为何种城市空间？这些问题不仅在设计界，同时在社会各界成为备受瞩目的话题焦点。

其次，在风景园林专业内，清溪垃圾填埋场再利用国际竞赛的中标方案在设计理论层面引起了业内的广泛讨论。2001年底，由美国景观设计公司James Corner Field Operations (JCFO) 带领的参赛团队所提交的竞赛方案"生机景观"（Lifescape）获得清溪公园国际竞赛的头奖。这是一个具有灵活性和适应性的设计方案。方案中提出了由线（threads）、面（mats）和岛（islands）所组成的网络矩阵，并以此最大化场地的可达性与流动性。"生机景观"描绘了随着时间推移分期建立场地植被与生态系统的策略，并同时兼顾为人们提供多种的活动空间。这种景观更多的受到时间与过程的影响，而非空间与形式(JCFO 2001)。用于塑造这个大尺度"生机景观"网络的三个最主要的技术手段是土壤改良、演替性植栽和地形塑造 (Corner 2005)。

此时，在美国风景园林领域的设计理论中，正值景观都市主义（Landscape Urbanism）成为讨论热点的时期，而该中奖方案的主设计师James Corner本人就是景观都市主义的主要旗手之一。清溪公园"生机景观"的方案也自此被专业人士和业内媒体广泛的作为景观都市主义的经典案例进行引用和讨论。在2006年出版的著作《景观都市主义读本》一书中，编者Charles Waldheim认为该方案显示了成熟的景观都市主义作品是如何将各种截然不同甚至可能相互矛盾的元素进行叠加和协调的。该作品典型的特征是具有多种详尽的分析图，包括分期开发、动物生境、植被演替、水文系统、功能布局和规划机制，这些分析表明了设计者对于大尺度项目所要面对的极度的复杂性的深刻理解 (Waldheim 2006)。"生机景观"被认为是一个弹性的方案，其表达了一种可以和多种影响因素进行对话的强有力的设计逻辑，可

One can only enter the fenced off landfill by appointment or with guided group tours (Fig.01-02). Only two small landscape projects are currently under construction, both on the perimeter of the site. The first is a 1 hectare playground (Fig.03-06), and the other is a sports park of about 8.5 hectares. So what are the reasons this landfill to park project has attracted so much attention?

2. A Hotly Debated Subject

First of all, the scale of Fresh Kills Landfill and the role it played over the years has put it into worldwide focus. Occupying an area of 890 hectares, Fresh Kills (Fig.07) was the world's largest sanitary waste landfill. On Sep. 5th, 2001, six months after Fresh Kills' closing announcement, the New York City municipal government announced an international competition for transforming the landfill into a public park. Only six days after the competition announcement, the September 11 event shocked the world. During the clearing of the 911 site, there was so much rubble and debris that the West Mound of Fresh Kills Landfill was reopened to serve as a sorting ground as well as a landfill site. The role that Fresh Kills Landfill played in connection with the 911 event truly put it in the world's attention and also changed peoples' sentiments towards it. The site was no longer just an unpleasant dump where garbage and unwanted items were buried; it became attached with the shock, grief, anger, and memories associated with the tragedy. Today, one can still see the white temporary structures (Fig.08) built for sorting the debris from the 911 site on the landfill grounds. One cannot visit this area without wondering what memories and emotions are buried here. What should be

图 07 从南区垃圾堆体顶部向北眺望清溪。
Fig 07 Panoramic view towards Fresh Kill from the South Mound.

图 08 用于清理从911事件地运来的废墟而搭建的白色临时构筑物。
Fig 08 White temporary structures built to sort debris from the September 11 ground.

以承受多种外界压力并在其之上生长壮大 (Czerniak 2007)。

再者，清溪垃圾填埋场的改造不光受到政府和设计界的关注，同时在设计过程中开展了大量的公众参与活动，引起了普通民众和多个非政府组织的广泛讨论。参与到该项目公众参与过程中的人员不仅包括场地周边以后每天会使用这个公园的居民（图 09），还包括曼哈顿和纽约市其他地区的各种非政府组织和兴趣团体，以及潜在的投资人士和集团。在其总体规划（Master Plan）阶段的公众参与过程就从 2004 年至 2006 年持续了 2 年之久。在这个过程中，公众经历了一个逐渐认识场地、了解场地、接受场地，然后才能畅想场地未来的过程。公众的意识的转变需要一段时间，不是一蹴而就的。在笔者与项目设计团队成员访谈时了解到，项目周边社区的很多居民在项目启动之初的情绪是抵触的，因为他们认为这么多年政府都在他们的“后院”倾倒垃圾，恶化了他们的生活环境，那么这次他们怎么能真的相信政府是要为他们谋福利呢？还有相当一部分公众对于垃圾场的安全性心存疑虑，" 我们到场地上来会得癌症吗？ "" 这里的垃圾会有核辐射吗？ " 这些疑问表达了公众内心对于垃圾场所产生的一种恐惧感，他们需要通过科学的引导才能对于垃圾场的内部结构、控制污染的环保措施以及未来的监测系统形成一个正确的认识。

为了促进和帮助公众转变对于垃圾场的固有认识，纽约市政府举办了各种宣传和研讨活动，努力和公众进行沟通，包括定期向公众开放的由专人带领并讲解的场地参观游，以及从 2010 年起已成功举办过三届的 " 潜入盛会 "（Sneak Peak）活动。每年秋季，在“潜入盛会”举办的这一天，清溪垃圾填埋场的一部分区域会向全社会开放（图 10），邀请大家“潜入”填埋场的内部来体验。人们可以在溪流中划皮划艇、爬上垃圾山的顶部眺望曼哈顿（图 11）、骑自行车、放风筝、参加关于堆肥和再利用的工作坊以及欣赏当地艺术家的作品等。最重要的是，这个活动为所有人提供了一次体验场地的机会，使大家认识到未来这个场地作为城市公共空间所具备的无限的可能性。关注清溪垃圾填埋场的公众在经过了这段长时间的参与后，对于垃圾场已经形成了新的认识，更加关注垃圾场及周边的生态建设与湿地恢复项目（图 12–15），他们开始憧憬垃圾场改造的未来，设想各种公园内适宜的活动项目。

the future role of such a site? What type of urban space could serve the public's needs while acknowledging the enormity of the tragedy? These questions were not only debated within the design profession but also by the public at large.

Within the field of landscape architecture, the winning scheme for Freshkills Park international competition raised much discourse in the realm of design theory. By the end of 2001, the competition entry entitled “Lifescape” by a team led by James Corner Field Operations (JCFO) was awarded first prize. Lifescape is designed to be flexible and adaptive. It proposes a matrix of threads, mats and islands to maximize opportunities for access and movement. The design scheme is “basically a strategy for revegetating the site over time, while outlining spaces where activities can occur.” This landscape is guided more by time and process than by space and form (JCFO 2001). The three primary techniques to help produce the large-scale matrix are soil making, successional planting and landform manipulation (Corner 2005).

Around this time, Landscape Urbanism was getting more and more attention in the theoretical discourse within the design realm, with James Corner being one of its main advocates. Subsequently, “Lifescape” became one of the most widely cited projects which demonstrate the design principles of Landscape Urbanism. In *the Landscape Urbanism Reader*, published in 2006, its editor Charles Waldheim argues that James Corner's project for Fresh Kills illustrates “mature works of landscape urbanism through ... accumulation and orchestration of absolutely diverse and potentially incongruous contents. Typical of this work ...

图 09 从南区垃圾堆体顶部向南眺望邻近的阿顿高地居住区。
Fig 09 View towards the adjacent Arden Heights neighborhood to the south from the South Mound.

图 10 不久前刚刚建成开放的清溪公园游客中心，其上进行了绿色屋顶的设计。
Fig10 The newly opened visitor center for Freshkills Park, which is equiped with a green roof.

are detailed diagrams of phasing, animal habitats, succession planting, and hydrological systems, as well as programmatic and planning regimes." "... they present an understanding of the enormous complexities confronting any work at this scale" (Waldheim 2006). Lifescape is considered to be a resilient scheme, which "presents a strong design logic that can sustain a dialog with multiple contexts, accommodating and growing from pressures put upon it" (Czerniak, 2007).

Besides attention from design professionals and the government, the transformation of Fresh Kills Landfill also involves the general public and multiple NGOs through an intense public participation process. Not only do people in the surrounding neighborhood (fig.09) of the landfill actively participate in the project discussions, but also many NGOs and interest groups and clubs in the region, as well as potential investors. As part of the master plan phase for Freshkills Park, a two-year long public participation process ran from 2004-2006. During this process, the public started to know the site, to understand the intention of the transformation, and to accept a general design vision. Eventually they began to imagine the future of the site together with the design team. A change in the public's attitude towards this type of project can only happen over time through education and involvement. It cannot happen overnight. At the beginning

图 11 清溪垃圾填埋场的至高点是远眺曼哈顿城市天际线的理想场所，在近处可以看到位于填埋场东北方向的湿地保护区。
Fig11 The highpoint in Fresh Kills Landfill is an ideal place to overlook Manhattan' s skyline, with a wetland reserve in the middle ground.

图 12 在清溪垃圾填埋场的范围内，人们已经观察到很多野生动物的活动。
Fig12 An abundance of wildlife has been observed within the limits of Fresh Kills Landfill.

图 13 在填埋场的栅栏上可以看到当地一所大学的研究人员安装的木制鸟窝以进行鸟类研究。
Fig 13 Wooden houses for birds are installed on the landfill fences by a local college to conduct study on birds.

图 14 美国新泽西州立罗格斯大学的一组研究人员在垃圾填埋堆体上种植了不同种类的树木来研究其根部是否会对于土层下的隔离层造成破坏。
Fig 14 A research team from The State University of New Jersey in Rutgers planted this group of trees to study if their roots would penetrate the landfill' s barrier layer underneath.

清溪公园的整个项目计划用30年的时间分期进行开发。这么长的周期是由多方面原因造成的，一方面，垃圾的降解和稳定化过程以及新的生态系统的建立需要时间。在垃圾填埋场封场以后，对于其渗滤液和填埋气体的收集、处理与监测要持续数年，一个填埋场往往要经过几十年的时间才能够达到最终稳定。另一方面，清溪公园项目需要逐渐的筹措建设资金，根据资金的可能性以及投资者的意愿进行局部区域的阶段性建设。

当然，从另一个角度来讲，这个项目周期可能太长了。在这个漫长的过程中，难免会出现项目组成员更替的问题，新的成员需要重新熟悉场地情况和项目的来龙去脉，老成员长期积累起来的知识和经验难免会部分遗失，政治经济上可能的变动没有人可以预期，项目之初的设计理念和愿景是否能一直贯彻到最后，没有人可以打保票。但是，有一点不可否认，就是这个漫长的过程使得这个重要的城市公共空间的塑造得以基于全社会所逐渐形成的共识之上，同时也使得这个污染场地的改造过程受到公众的监督，其既不是某个领导的长官意志，也不是某个精英设计师的纯自我表现。在社会各界参与的过程中，每个人都逐渐形成了对这个场地的认同感和归属感，这种无形的力量在形成社会凝聚力和城市活力上是不容忽视的。

清溪垃圾填埋场的改造之所以如此的备受关注，并不是因为其设计的空间形态多么的令人意想不到，也不是因为其华丽转身是多么的剧烈快速，而是因为它成为了一个热议的话题，引起了社会的、生态的、伦理的和设计理论的等多方面的讨论，挑战了传统的固有认识，既包括对于垃圾填埋场固有认识的挑战，也包括对于风景园林设计理念与思路的固有认识的挑战。

3. 多给点时间

景观设计一直就是建立在科学技术与人文艺术相结合的基础之上的。在垃圾填埋场再生这种特殊的项目类型上，工程技术方面所提出的要求更为严苛，因为其关乎环境污染问题，涉及到维护广大使用者的安全和健康，这就对于风景园林师提出了更高的要求，如何在这些限制条件下进行创造性的空间设计成为新的挑战（图16-22）。

在公众的认识上，对于垃圾填埋场的再利用需要一个转变的过程，因为长期以来垃圾场都是周边的祸害，又脏又臭，周边居民往往感到愤怒、无奈、甚至羞耻。在垃圾场周边居住的人群中很多是弱势群体和边缘群体，包括大量以捡拾垃圾为生的拾荒者。如何让公众接受、

of the Freshkills Park project, many people in the adjacent neighborhood were suspicious about it. They felt that they have been dumped upon in their "backyards" for so many years, that the government could not be trusted to do something positive. Some people were concerned with the safety of the landfill. "Will we get cancer from participating in a group tour of the landfill?" "Is the site radioactive?" These questions expressed the typical fears the public has for landfill reuse projects. These fears can only be allayed through appropriate guidance and education on the subject of the landfill's engineering structure, environmental controls and monitoring measures.

To assist the general public in adjusting their attitude towards the site, the municipal government held a wide-range of public meetings, events and seminars, as well as guided bus tours in the warm seasons. Since 2010, an annual public event named Sneak Peak has been successfully held on the site. On the day of the event, part of the landfill is completely open to the general public (Fig.10) and whoever is interested can come, observe, and experience the transformation. One can view Manhattan's skyline from the top of the landfill mound (Fig.11), kayak, ride bicycles, fly a kite, join composting and reuse workshop, or enjoy artworks by local artists. Most importantly, this event provides an opportunity for people to see with their own eyes the endless potential Fresh Kills Landfill offers as an urban open space. Through these many years of participation and discussion, the public have established a new vision for the landfill, and begun to see the potential that this huge park can offer (Fig.12-15).

The entire project is expected to be completed over a thirty year period via several different phases. Many factors contribute to this long time frame. The garbage degradation process can

图15 清溪从垃圾填埋场中间蜿蜒的流过。只有当看到地面上伸出的填埋气体收集管道时，人们才恍然意识到自己其实是站在成千上万吨的垃圾之上的。
Fig 15 Fresh Kill flows through the landfill site. Only the protruding methane pipes remind you that you are standing right on top of tons of garbage.

图 16–18 老式的填埋气体收集及监测设施，可以看到它们高出地平面很多并且破坏了场地地形的整体感和连贯性。
Fig 16–18 Old fashioned infrastructure to collect and monitor methane generated by the buried waste.They are sticking out from the ground and interrupting the landform.

图 19–20 位于东区垃圾堆体的新式检修口与地平面基本齐平，更好的融入到整体的景观地形中。这些检修口如此设计是在得知清溪垃圾填埋场将在未来作为城市公园之后。
Fig19–20 New manholes on the East Mound which is flush with the ground and merge into the landscape. These manholes were designed after the decision that Fresh Kills Landfill will be turned into a parkland in the future.

图 21–22 位于东区垃圾堆体上的一处雨洪导流槽，雨水最终被引入低处的蓄水池中。
Fig21–22 A downchute on the East Mound designed for storm water run–off which is guided to detention basins in the lower spot.

21

22

图 23 清溪垃圾填埋场的南区垃圾堆体在 20 世纪 90 年代就已经封场，现在整个垃圾堆体已在自然演替中被植被所覆盖。
Fig 23 The South Mound of Fresh Kills Landfill was closed in the 1990s, which is covered by vegetation now through natural succession.

理解并监督这个转变过程是很重要的，例如清溪垃圾填埋场项目通过先在场地中进行小规模的项目建设及组织参观体验活动，为广大公众提供与场地接触以加深了解的机会，使他们慢慢地来认识、理解和支持这种转变。垃圾填埋场的环境污染治理与监测是一个长期的过程，自然生态系统的建立是一个长期的过程（图 23），公众对于垃圾场封场后功能转变的认识和接受也需要一个过程。因此，在这类项目中应该允许相对较长的项目周期，而不应急于求成。某些情况下，这类项目是在强烈的政治诉求下被推动的，这在中外都是非常正常的现象，但切不可一味地为了快速树立政绩工程而无视这类项目特有的客观周期规律和形成社会认同感的时间需求。中国很多地方政府仍然把 " 快 " 作为重大项目的突出成绩之一是值得商榷的。□

注：本文全部照片由郑晓笛拍摄

参考文献

Corner, J. (2005). "Lifescape - Fresh Kills Parkland." Topos 51: 14-21.
Czerniak, J. (2007). Legibility and Resilience. Large Parks. J. Czerniak and G. Hargreaves. New York, Princeton Architectural Press: 215-251.
JCFO (2001). Lifescape (Competition Narrative). New York, Field Operations.
Waldheim, C. (2006). Landscape as Urbanism. The Landscape Urbanism Reader. C. Waldheim. New York, Princeton Architectural Press: 36-53.

take decades for the landfill to eventually reach a stabilized state. Monitoring measures for leachate and methane will last for years to ensure that it is within safety limits. As well, funding for the park can only be acquired incrementally based on fiscal availability and investor interest. Thus the park will be developed sequentially in small portions.

However, it might be true that this planned duration is a bit too long. Potential problems with this long time span include the inevitable change of project members. New members need to pick up all the complications and histories of past decisions and actions, and accumulated knowledge and experience might get lost during this process. No one can predict the political and economic condition years from now, and no one can guarantee that the initial vision and design concept can be carried all the way through. While this long time span allows for many potential problems and obstacles to impede the park's development, it must be acknowledged that it allows for a development process based on societal consensus and public scrutiny. The project is neither a pure political pursuit driven by one government official, nor a mere expression of one designer's ego. This type of intense public participation is essential for the creation of social vitality in a community. Everyone involved starts to feel an ownership of the site, as well as a sense of belonging.

The Fresh Kills project has attracted so much attention over the past decade not because of any unprecedented spatial design or a drastic makeover, but rather because of the wide range of topics it addresses. Social, ecological, ethical, and theoretical issues have been examined. This project challenges our traditional way of thinking on many dimensions, including the potential role a closed landfill can play in an urban setting, and how landscape architects can approach projects of such complexity and scale.

3. Give Some Time

Landscape architecture has always been a discipline that bridges science, technology and the arts. In tackling this specific type of landfill regeneration projects, design professionals are required to cope with environmental engineering constraints based on public health and safety. How to generate attractive public open spaces under such constraints becomes a challenge for landscape architects and calls for creative thinking (Fig.16-22).

It takes time for the general public to understand and accept the transformation and reuse of closed landfills. Neighboring residents have suffered from their negative environmental and economic impact, and experienced anger, frustration and shame. Many people living close to a landfill are disadvantaged groups. It is important to have multiple forms of public participation to help people realize the potential of a closed landfill after its reclamation, to appreciate the opportunities it has to offer, and to scrutinize the transformation process. In Fresh Kills Landfill, such efforts were made through parceled sequential development and public events such as Sneak Peak. It is only by witnessing and participating in the development process that the public can become advocates for this regeneration effort. In landfill regeneration projects, both environmental reclamation (Fig.23) and public acceptance require a period of years, or even decades. In some cases, landfill regeneration projects are initiated by strong political will, which is common both in China and abroad. However, it is wrong to compress project cycles to extremely short periods of time merely to showcase political achievement. Time must be allotted for natural processes and the formation of social identity. It is debatable whether local governments in China should still cite "fast" as an outstanding achievements in the development of such projects.■

Reference

Corner, J. (2005). "Lifescape - Fresh Kills Parkland." Topos 51: 14-21.
Czerniak, J. (2007). Legibility and Resilience. Large Parks. J. Czerniak and G. Hargreaves. New York, Princeton Architectural Press: 215-251.
JCFO (2001). Lifescape (Competition Narrative). New York, Field Operations.
Waldheim, C. (2006). Landscape as Urbanism. The Landscape Urbanism Reader. C. Waldheim. New York, Princeton Architectural Press: 36-53.

作者简介：

郑晓笛 / 女 / 美国注册风景园林师（宾夕法尼亚州）/ 清华大学建筑学院博士在读 / 中国花卉园艺与园林绿化行业协会国际部主任

Biography:

Xiaodi Zheng / Female / Registered Landscape Architect in PA, USA / Doctoral Candidate at School of Architecture in Tsinghua University / Director of International Affairs,Chinese Flowers Gardening and Landscaping Industry Association

查尔斯 · 沙（校订）
English reviewed by Charles Sands

图 01 鸟瞰效果
Fig 01 The Birds' Eye View

鹭翔莲影

Dances with Birds on Lotus.

澳门氹仔城市湿地公园规划设计

LANDSCAPE DESIGN FOR THE FIRST WETLAND PARK IN MACAU.

中文标题：鹭翔莲影——澳门氹仔城市湿地公园规划设计
组　别：设计作品组
作　者：魏忆凭 金 英 马晓宾
指导教师：李 敏
学　校：华南农业大学林学院
学科专业：风景园林硕士研究生
研究方向：风景园林规划与设计
分　类：2012“园冶杯”国际竞赛设计作品组类作品 一等奖

Title: Dances with Birds on Lotus. Landscape Design for the First Wetland Park in Macau
Degree: Design Group
Author: Yiping Wei, Ying Jin, Xiaobin Ma
Instructor: Min Li
University: College of Forestry, South China Agriculture University
Specialized Subject: Master of Landscape Architecture
Research: The Planning and Design of Landscape
Category: The Second Prize in Design Group of 2012 "YuanYe Award" International Architecture Graduation Project/Thesis Competition

图 02 场地现状照片
Fig 02 The Photo of the Site

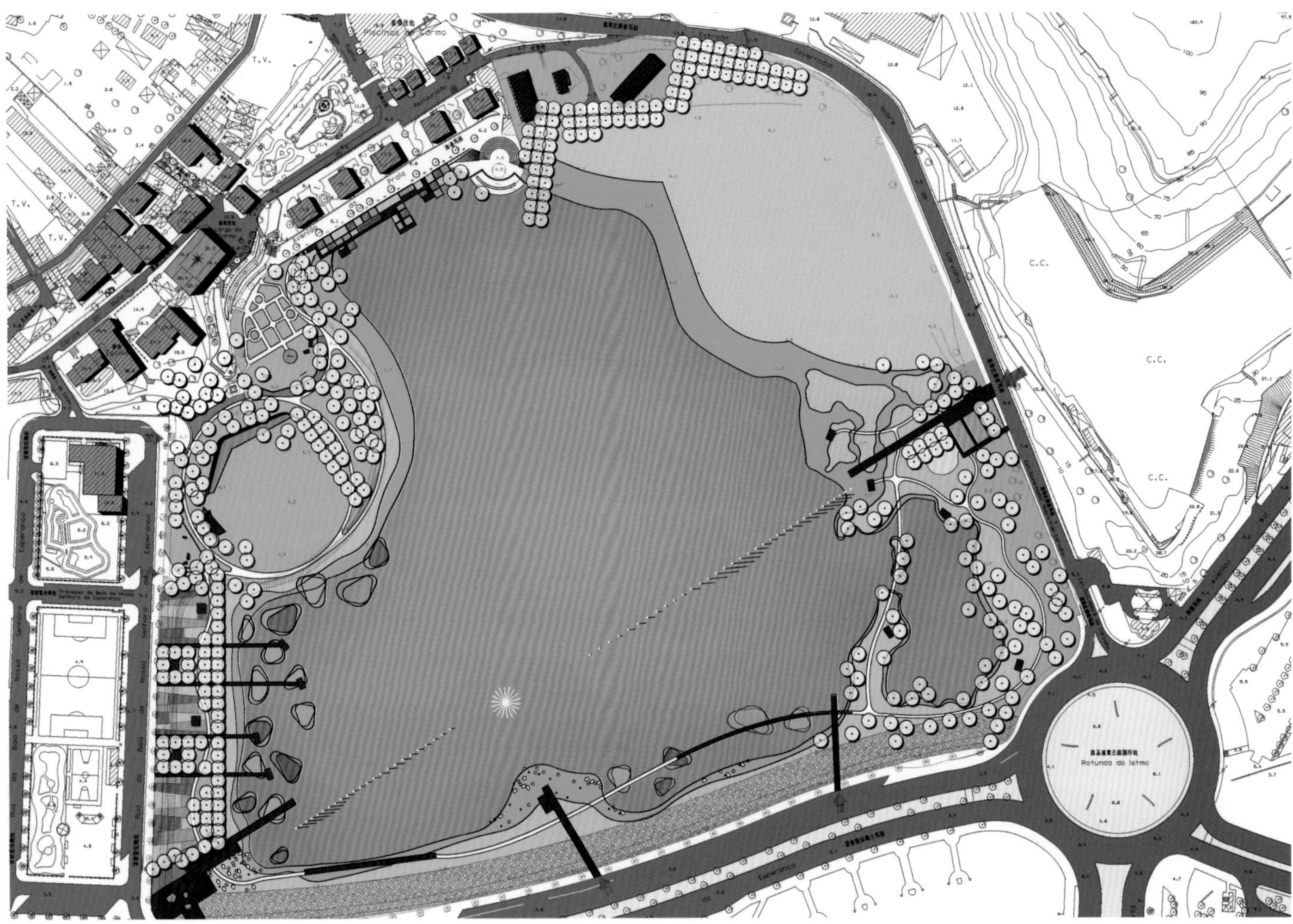

图 03 总平面图
Fig 03 The Master Plan

作品简介：

本项目场地位于澳门氹仔岛，历史上原是一片海滨滩涂，澳门持续不断的大规模填海运动，最终使其成为了一个被高楼围绕的水塘(图01-02)。由于城市经济发展的压力，这块土地极有可能被用作赌场扩建或交通枢纽的建设用地，而这些都不是澳门市民所希望的。作为尊重土地自然属性的景观设计师，我们和澳门民政总署园林绿化部的专业人员达成共识：应在该场地上建成澳门的第一个城市湿地公园。

澳门古称“莲岛”，莲花也是澳门区徽的标志图案。公园场地内有片原生的红树林湿地，是澳门城区仅存的野生鹭鸟栖息地。洁白的鹭鸟每天在湖面上盘旋飞翔，吸引了许多游人驻足观赏。因此，设计中提炼了莲花与鹭鸟形象，构成场地中最突出的空间构图要素(图03)。

本设计主题创意为“鹭翔莲影”。因为在中国传统文化里，鹭鸟与莲花通常是成对出现的吉祥物，许多诗词、瓷器中，常有“鹭莲戏水”的画面，表达“鹭莲相亲”的美好意境。公园标志性景观为湖中一列起伏变化的景观灯柱，白天可供鹭鸟歇脚，夜晚则仿佛光影诗篇。该柱列好似鹭鸟飞翔的空间轨迹，又像鸟儿欢唱跳跃的音谱。在柱列的中心位置，设计了一朵洁白圣洁的莲花，突出“鹭翔莲影”的意境(图04-15)。

Introduction:

The Site is located on Taipa island of Macau which was a seaside shoal, but currently is surrounded by massive tall buildings made possible by coastal reclamation (Fig 01-02). Due to the pressure of the city's economic development, the land was

图 04 主入口观景平台
Fig 04 Main Entrance Timber Deck

图 05 滨湖栈道
Fig 05 Timber Road

proposed to be used for a casino expansion or construction land transportation hub, neither of which is desired by the majority of people in Macao. Therefore, the government and the public have come to a mutual consensus to build the first urban wetland park in Macau.

Macau was originally known as 'Lotus Island' and the lotus flower is used on the current emblem of Macau. The existing site consists of native mangrove wetlands, which are a sanctuary for the city's remaining wild egrets. The white egrets hovering over the lake are a popular attraction for visitors. Therefore, the design takes the image of egrets combined with the lotus flower, to constitute the essence of the most prominent compositional elements (Fig 03).

The design theme is "Lu Xiang Lin Ying", which means egrets soar over the shadow of the lotus. In traditional Chinese poems and porcelain, egrets and lotus often appear in pairs. The iconic feature of the park is a fluctuating light fixture in the middle of the lake. It provides a resting place for the birds during the day and at night becomes a luminous sculpture. The column-like orientation imitates the trajectories of the flying Egrets and their playful movement in the water. In the middle of the columns, a white lotus sculpture has been placed to strengthen the meaning of "Lu Xiang Lin Ying" (Fig 04-15).

查尔斯·沙（校订）
English reviewed by Charles Sands

图 06 观鸟屋
Fig 06 Birds Observatory

图 07 立体花园
Fig 07 Hanging Garden

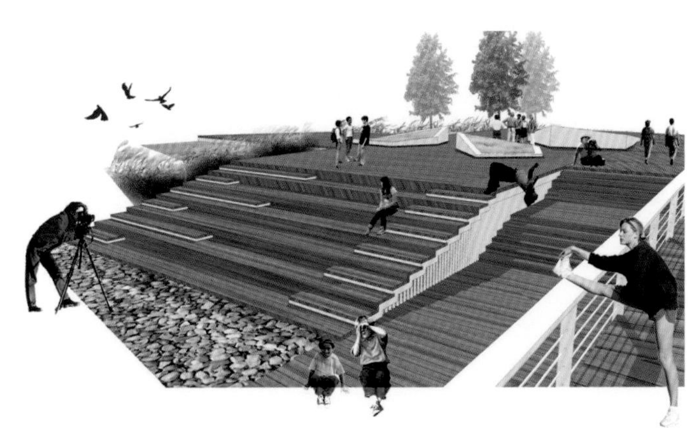

图 08 台阶看台
Fig 08 Main Entrance Bleacher

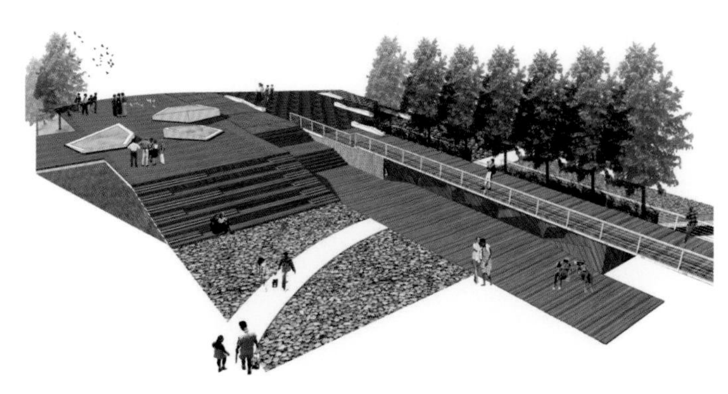

图 09 主入口广场
Fig 09 Main Entrance Plaza

图 10 观景盒子
Fig 10 Timber Boxes

图 11 观景盒子中的视野
Fig 11 The View of Timber Boxes

图 12 生态小岛
Fig 12 Eco Islands

图 13 休闲草坪
Fig 13 Lawn

图 14 都市园艺
Fig 14 Urban Garden

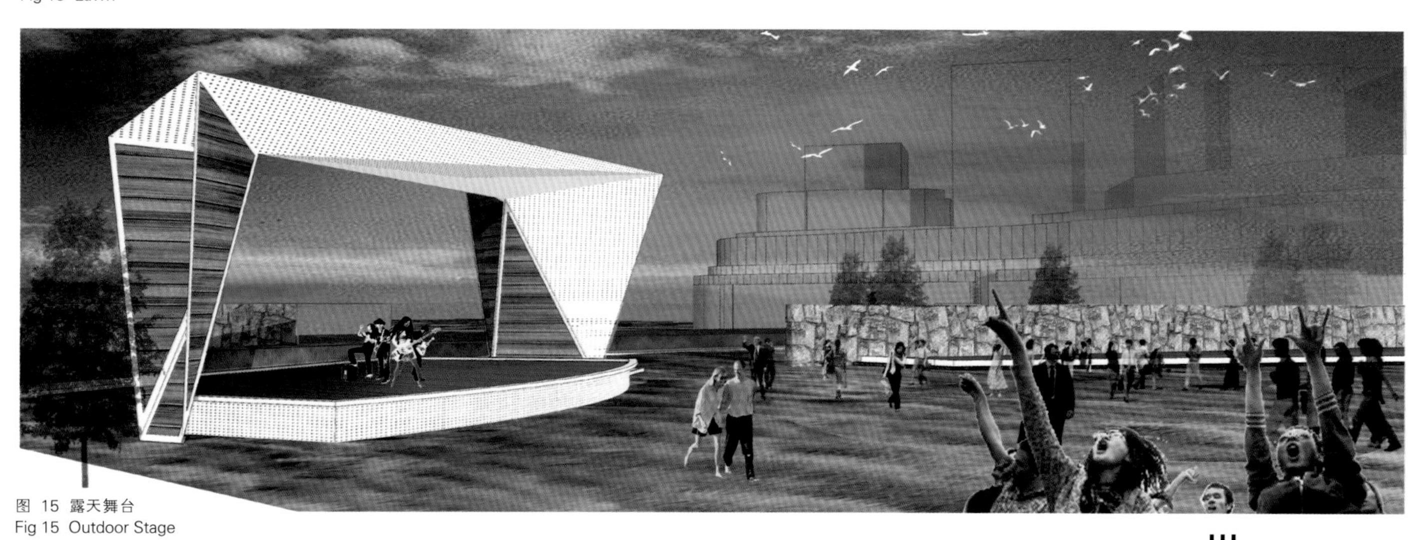

图 15 露天舞台
Fig 15 Outdoor Stage

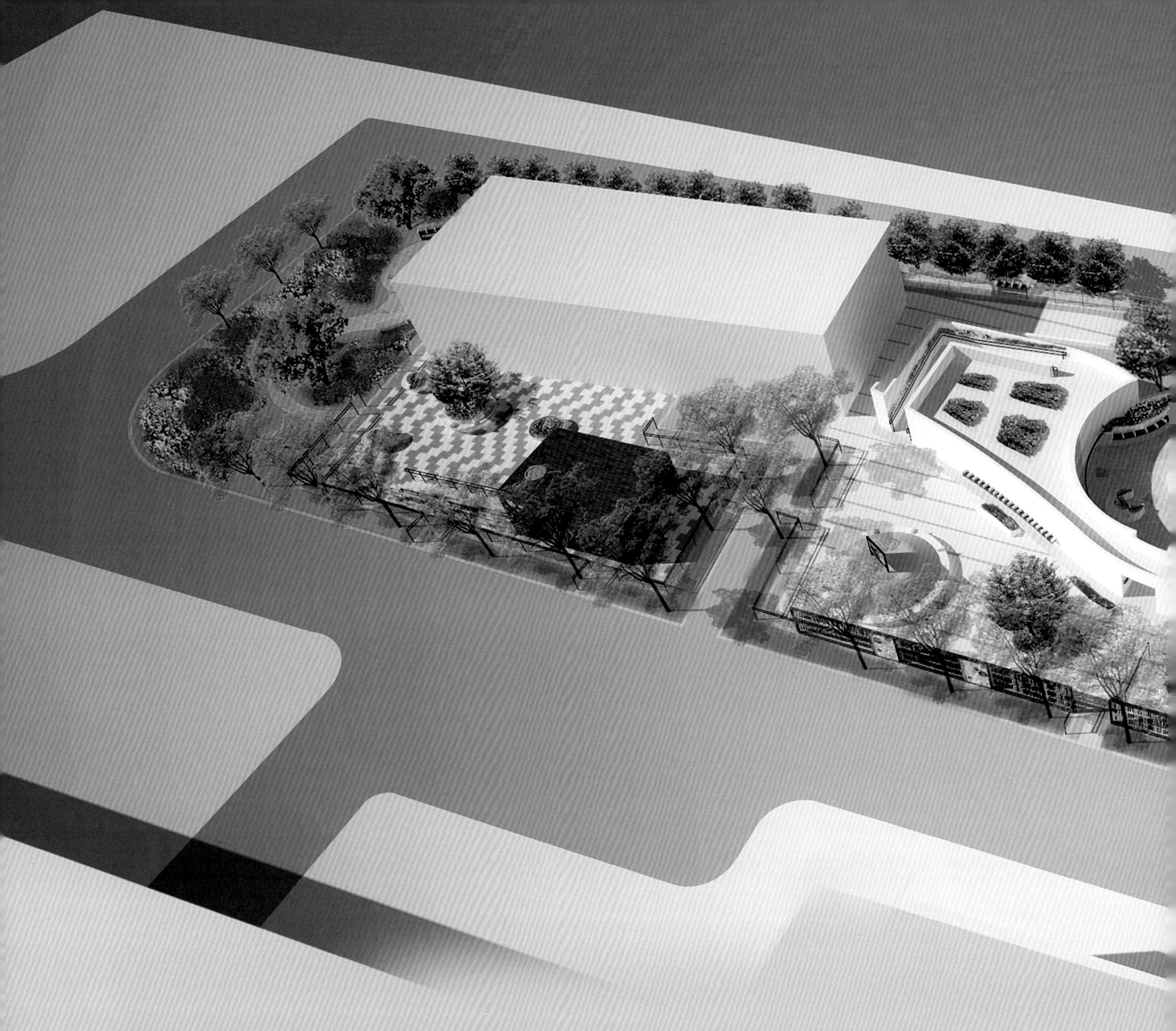

无障碍花园

BARRIER-FREE GARDEN

合肥残疾人托养中心户外康复空间设计

HEFEI DISABLED REHABILITATION CARE CENTRE OUTDOOR SPACE DESIGN

中文标题：无障碍花园——合肥残疾人托养中心户外康复空间设计
组　　别：设计组
作　　者：乔　方
指导教师：李一霏
学　　校：武汉科技大学
学科专业：艺术设计
研究方向：环境艺术方向
分　　类：2012“园冶杯”国际竞赛设计组类作品二等奖

Title: Barrier-Free Garden -- HEFEI DISABLED REHABILITATION CARE CENTER OUTDOOR SPACE DESIGN
Degree: Design Group
Author: Fang Qiao
Instructor: Yifei Li
University: Wuhan University of Science and Technology
Specialized Subject: Art Design
Research: Environment Art
Category: The Second Prize in Design Group of 2012 "YuanYe Award" International Architecture Graduation Project/Thesis Competition

图 01 鸟瞰
Fig 01 Aerial view

1 2 3 4

图 02 区位图（1）+（2）
Fig 02 Location（1）+（2）

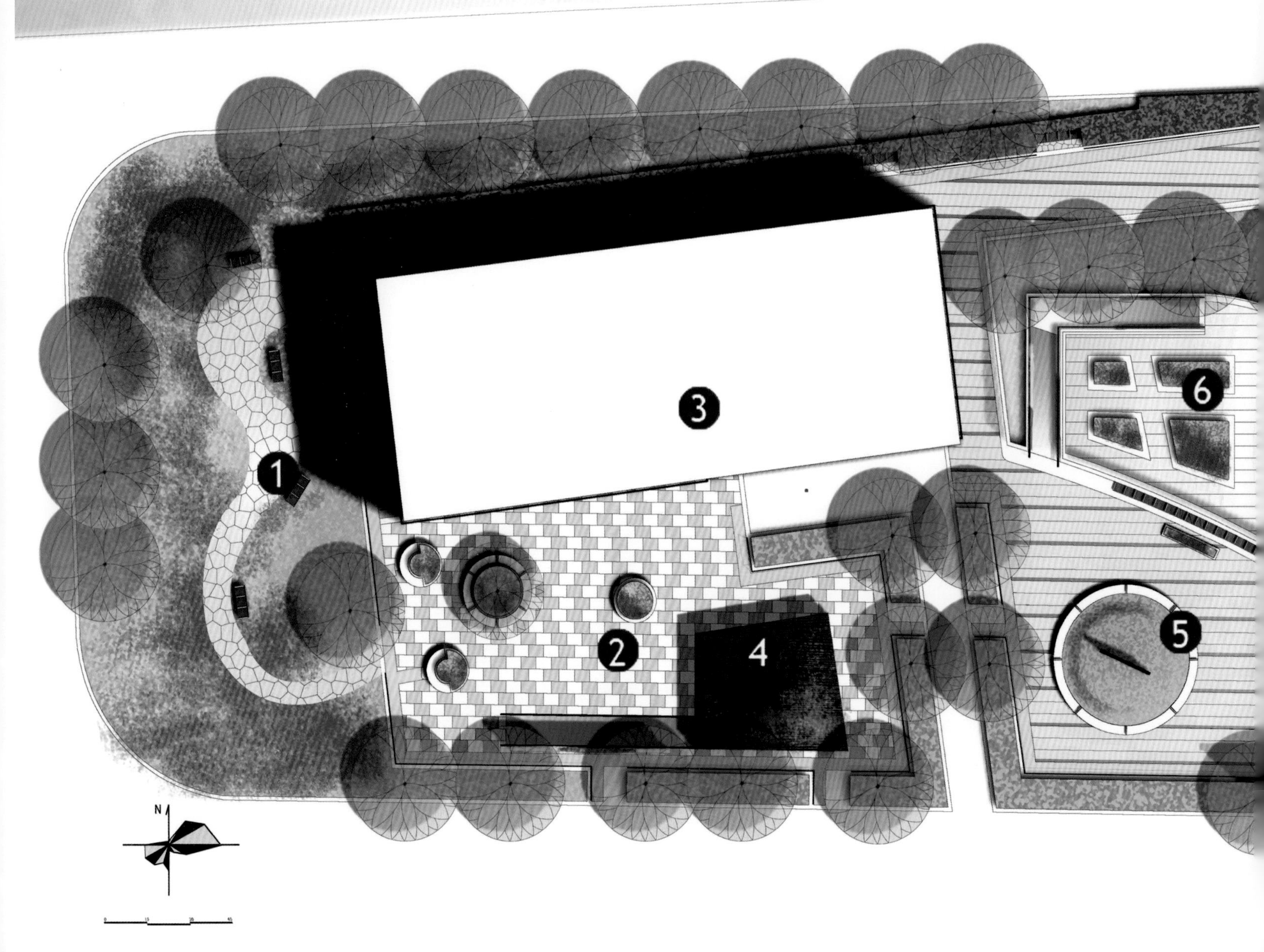

图 03 总平面图
Fig 03 Plan

❶ 园艺花园
❷ 左广场
❸ 食堂建筑
❹ 绿屋
❺ 微笑显屏
❻ 二层花坛
❼ 休息区
❽ 活动广场
❾ 二层休息区
❿ 黑板交流区
⓫ 转弯训练 / 交流区
⓬ 许愿树

作品简介:

该设计是合肥市残疾人托养中心户外康复空间景观（图 01），位于安徽省合肥市瑶海区磨店乡大众路（图 02–06）。该中心建筑面积 5076m²，设置托养床位 100 张。该设计本着为残疾人提供良好的承载物理治疗、进行心理康复的户外环境为目标，以人文理念为出发点，将园艺活动、交流、物理锻炼、安静休息等功能组织起来（图 07–10）。设计采用曲线等具有生命力而活泼的图形符号，以清新淡雅的色调为背景（图 11），创造出让残疾人轻松舒适的环境（图 12–14）。

Introduction:

The project creates an outdoor space for a disabled rehabilitation center (Fig.01). The site is located by a Public road in Modian village, Yaohai District, Hefei City, Anhui Province (Fig.02-06). The center is 5076 square meters and contains100 bed units. The design intent is to provide patients with a suitable outdoor environment for physical therapy and psychological rehabilitation. Beginning with a humanistic design approach, the project organizes the related functions, such as gardening, and physical exercise (Fig.07-10). The design utilizes curves and others vivid elements. In combination with a solid colored background (Fig.11) creating a relaxing and comfortable environment (Fig.12-14).

查尔斯 · 沙（校订）
English reviewed by Charles Sands

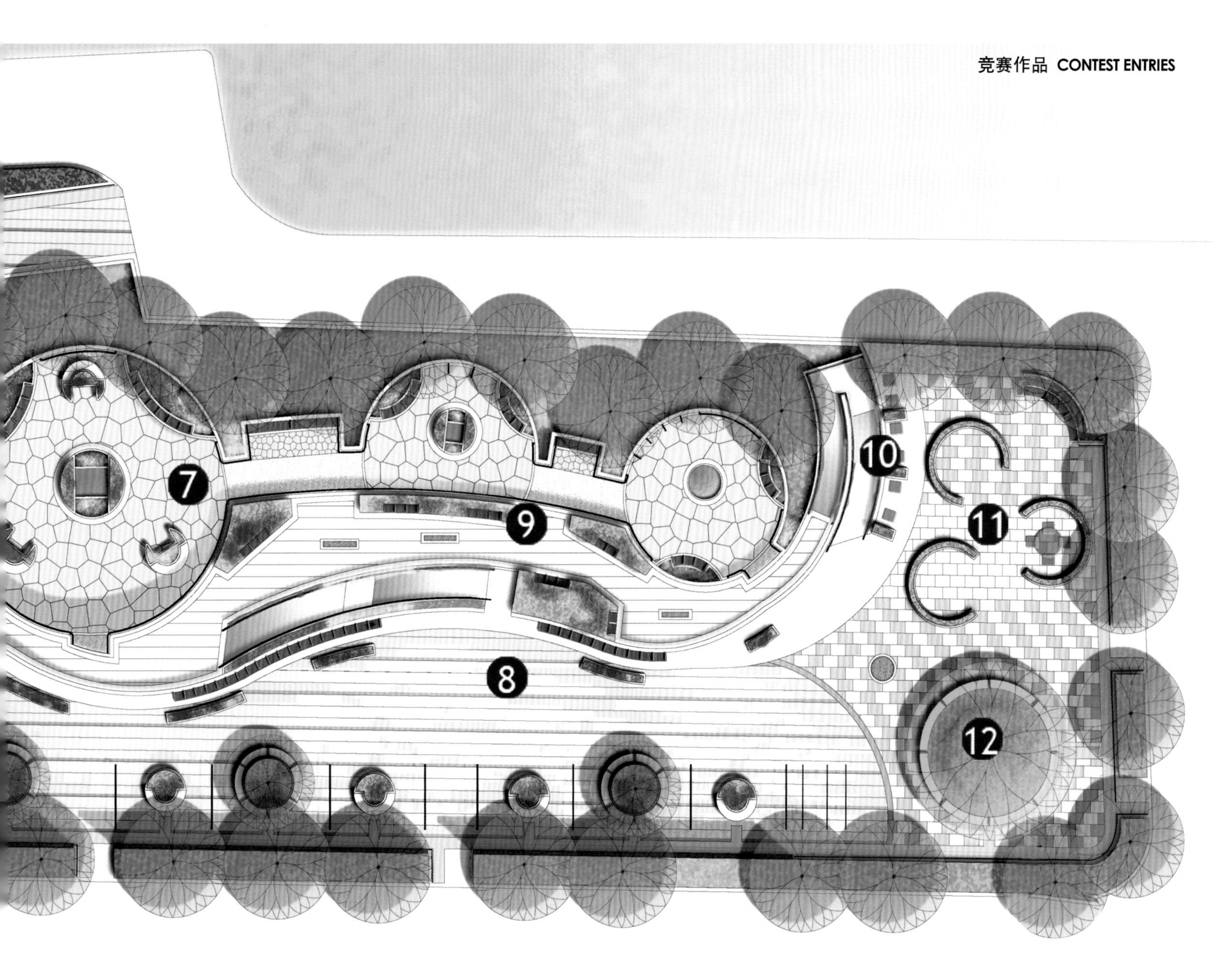

图 04 现状（1）
Fig 04 Original scene(1)

图 05 现状（2）
Fig 05 Original scene(2)

图 06 现状（3）
Fig 06 Original scene(3)

图 07 锻炼区
Fig 07 Exercise district

图 08 园艺区
Fig 08 Garden district

图 09 交流区
Fig 09 Communication district

图 10 休息区
Fig 10 Rest district

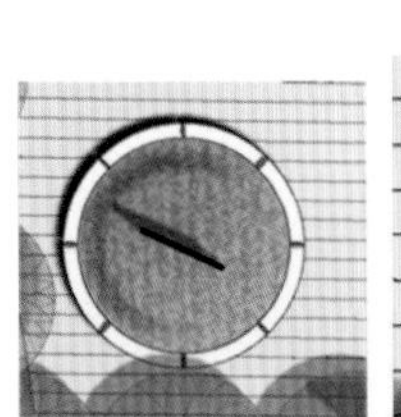

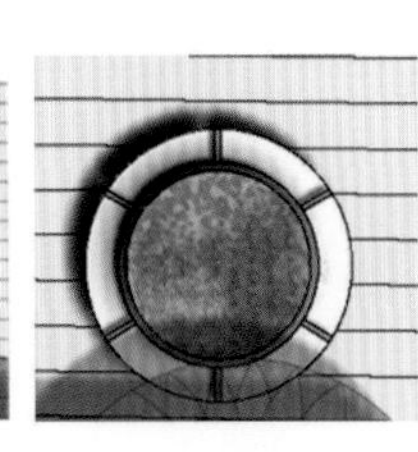

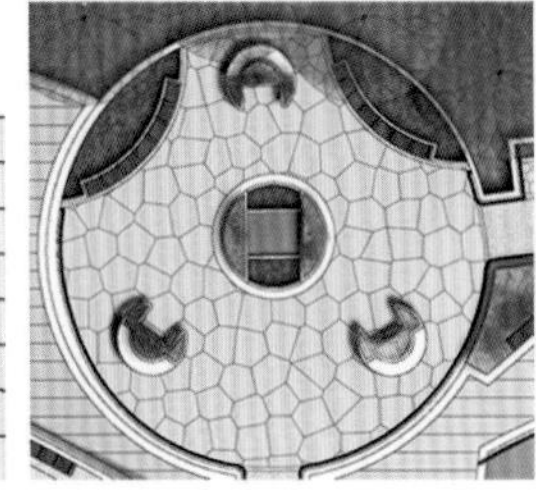

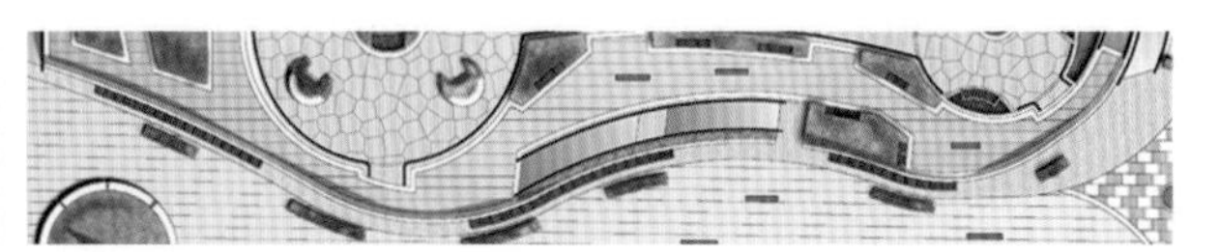

图 11 圆形和曲线元素
Fig 11 Circular and curve element

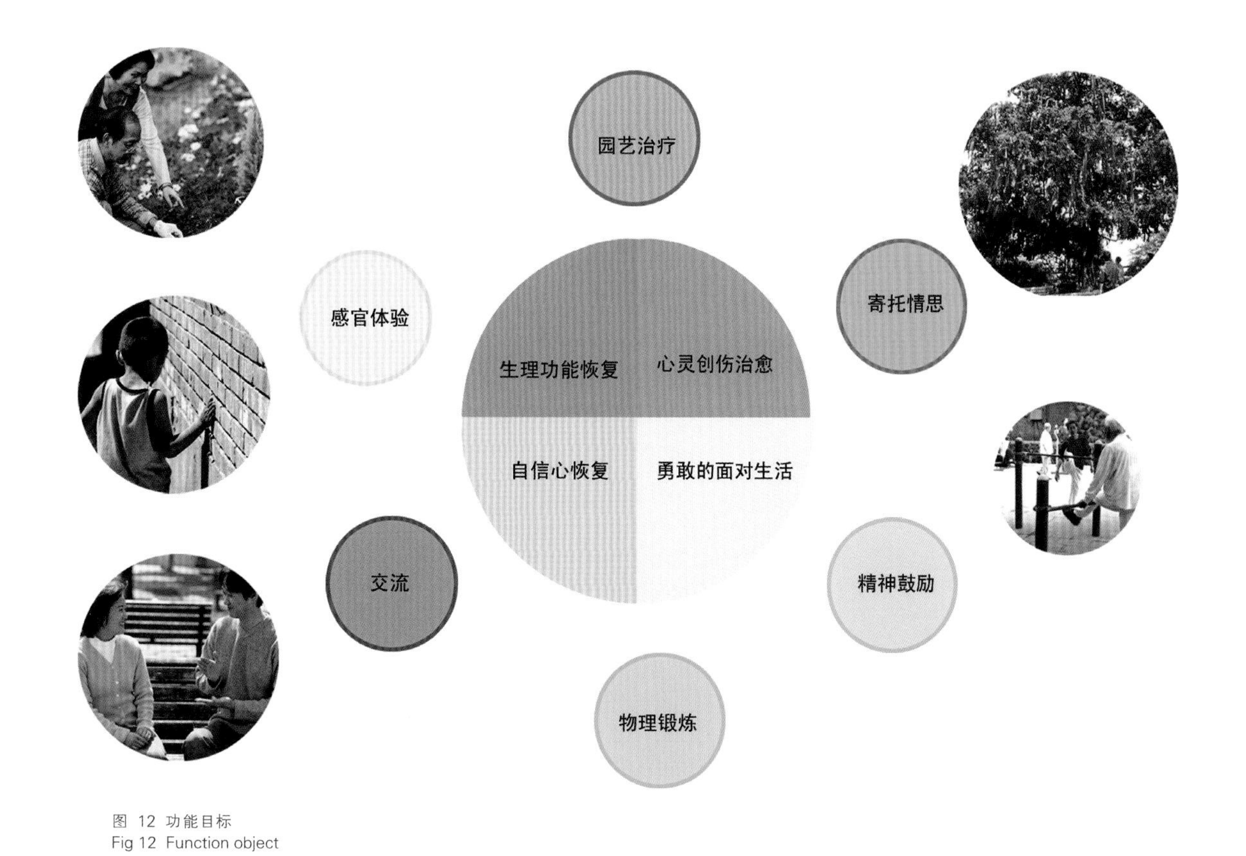

图 12 功能目标
Fig 12 Function object

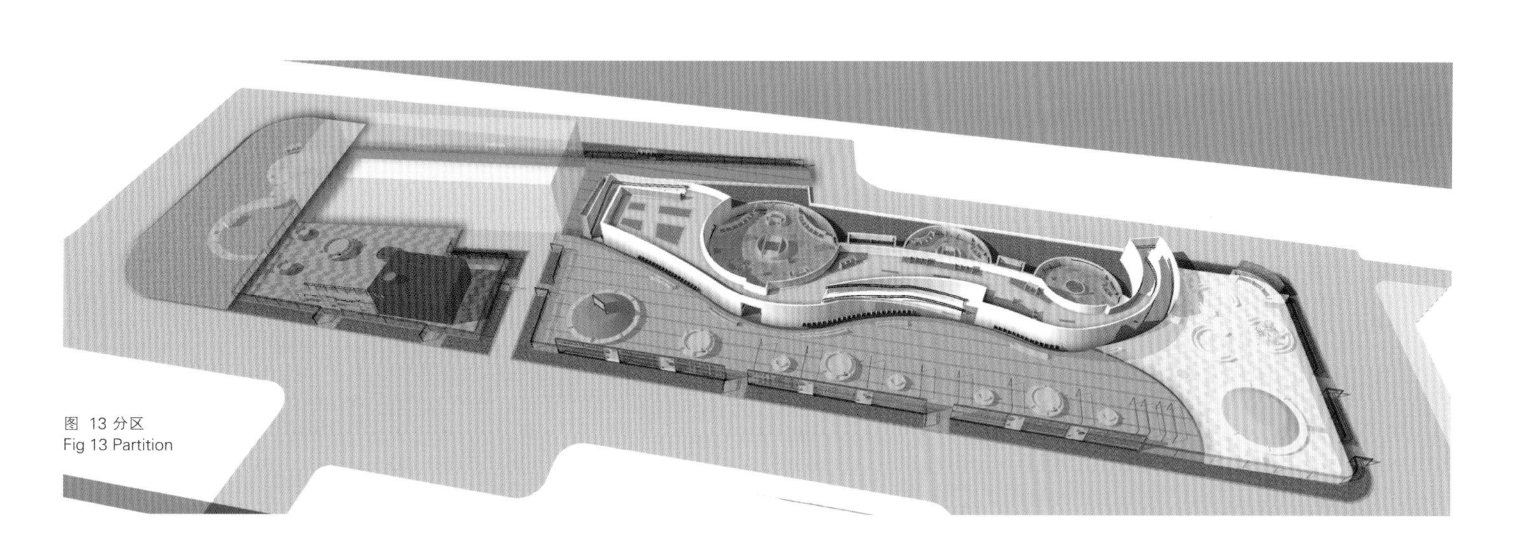

图 13 分区
Fig 13 Partition

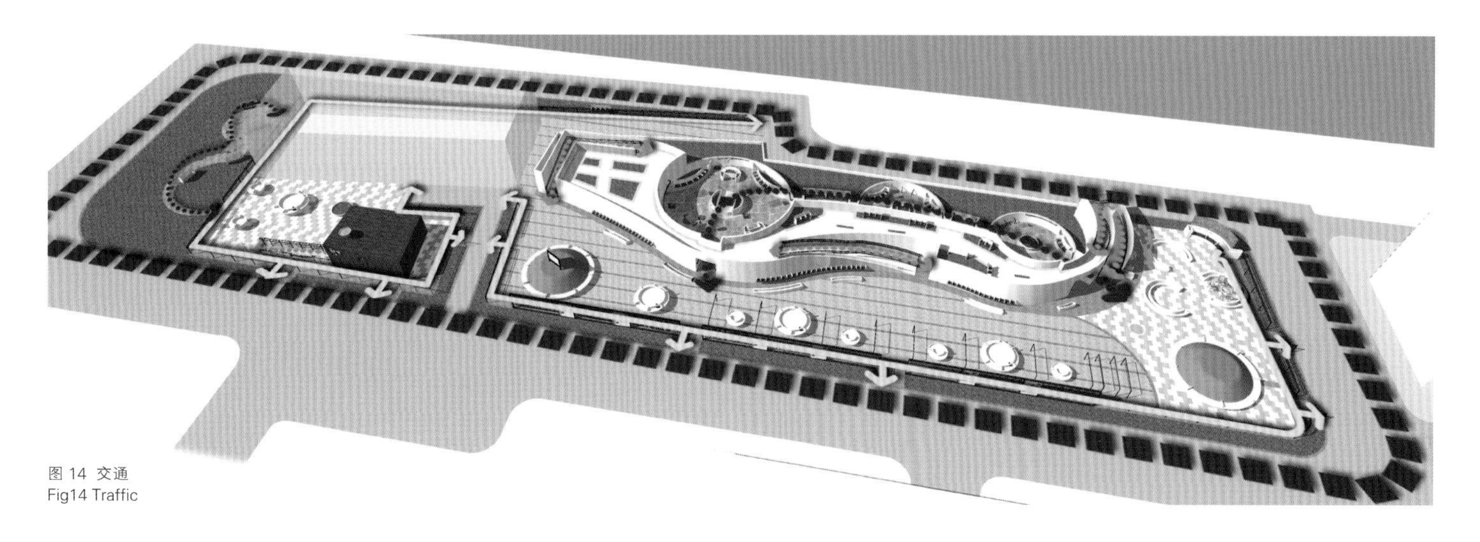

图 14 交通
Fig14 Traffic

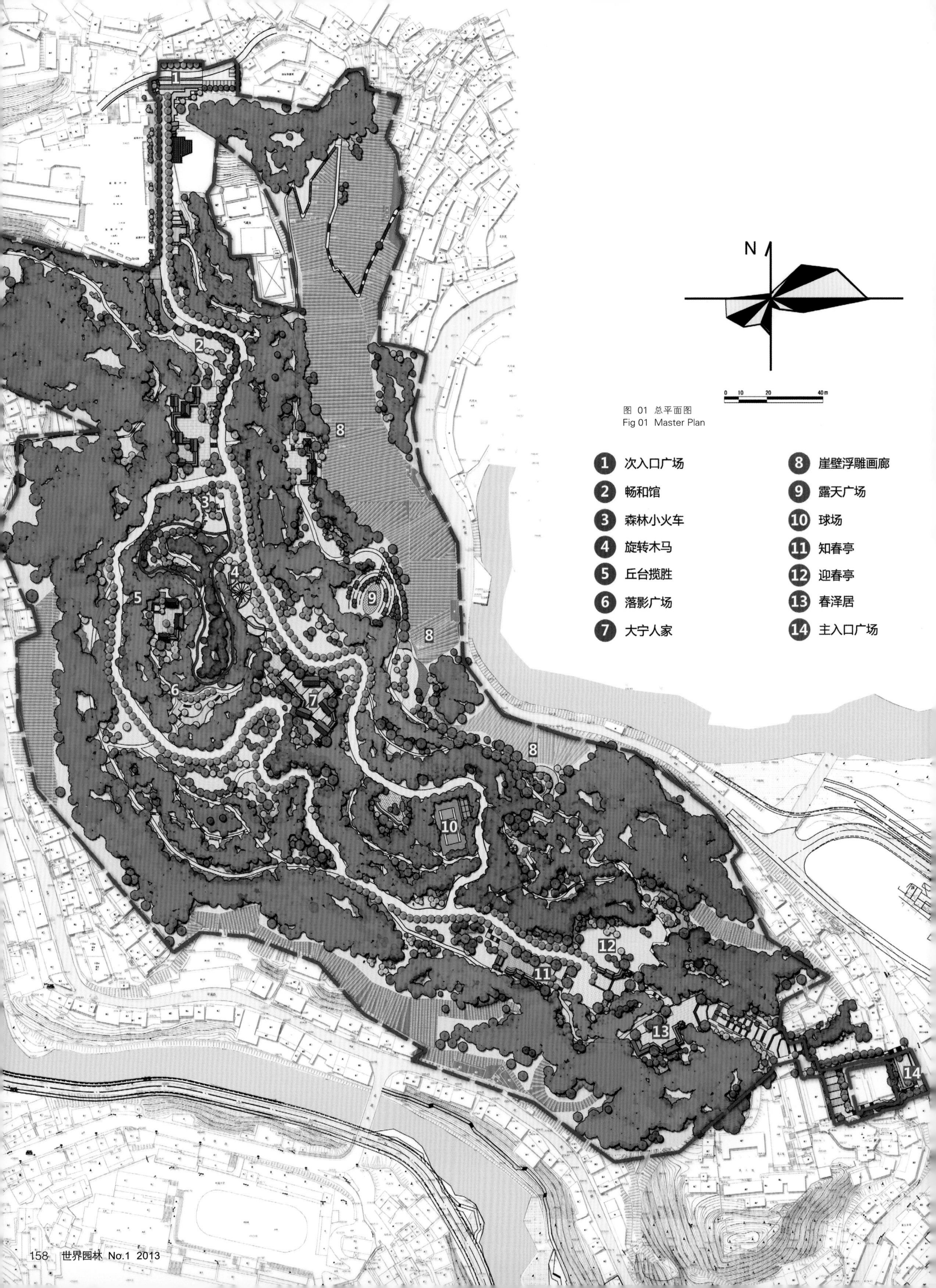

图 01 总平面图
Fig 01 Master Plan

图 02 次入口断面图
Fig 02 Secondary Entrance Section

百里画廊 万峰叠美
HUNDRED MILE MOUNTAIN GALLERY
——重庆巫溪城市综合性山地公园景观规划设计
LANDSCAPE DESIGN OF THE CITY'S COMPREHENSIVE MOUNTAIN PARK IN WUXI, CHONGQING

中文标题：百里画廊万峰叠美——重庆巫溪城市综合性山地公园景观规划设计
组　别：设计
作　者：骆　畅　徐思瑶　何虹利
指导教师：周建华
学　校：西南大学
学科专业：城市规划
研究方向：风景园林方向
分　类：2012“园冶杯”国际竞赛设计类作品二等奖

Title: Hundred Mile Mountain Gallery -- Landscape Design of the City's Comprehensive Mountain Park in Wuxi, Chongqing
Degree: Design Group
Author: Chang Luo, Siyao Xu, Hongli He
Instructor: Jianhua Zhou
University: Southwest University
Specialized Subject: Unban Planning
Research: Landscape
Category: The Second Prize in Design Group of 2012 "YuanYe Award" International Architecture Graduation Project/Thesis Competition

图 03 次入口景观效果图
Fig 03 Secondary Entrance Perspective

作品简介：

本设计位于重庆巫溪县，区位优势明显，占地面积较大，交通较为便利，现有植被基础较好。综合分析用地现状，将龙头山公园定性为：以良好的生态环境为基础，集休闲游憩、康体健身、文化娱乐等功能为一体的开放型区域性城市综合公园。本方案利用崖壁的特色地形形态，基于巫溪的文化，营造出“百里画廊”特色景观，

同时根据山地地势形成丰富的视觉空间变化、富有层次的山地植物群落景观以及高低错落、或私密或开敞的活动场地，并将当地的独有文化引入到场所中，普及历史文化的科普教育，充分体现场所精神，最终打造成集休闲游憩、康体健身、文化娱乐等功能为一体的开放型区域性城市综合公园（图 01）。

在公园的布局上，结合使用人群的不同活动爱好，和山地的地形空间，将中老年人的运动器材和场地安排在地形较低处，将儿童游耍的场地和设施安排离入口较近的位置。园区内将道路分为三级，为车流和人流提供便利的交通。根据适地适树原则，合理布局植物分区，形成四季景观的丰富变化。合理利用场地现有资源条件，结合其山地地形，使其成为一个能够供市民活动游憩的城市综合公园（图 02–08）。

Introduction:

The Wuxi Comprehensive Mountain Park is located in Wuxi County, Chongqing. The site has an excellect geographical location being easily accessible and having an extensive coverage of existing vegetation.

On the basis of urban land use, the park will be qualified as an open regional comprehensive park with multiple functions such as leisure recreation, sports fitness, and cultural entertainment.

The design takes advantage of the distinctive terrain to showcase the characteristic scenery of the River Wu in the Wuxi region.

Variable spaces with richly layered highland flora provide activity spaces at random heights and distribution according to the mountainous terrain. The local culture is also introduced, along with scientific and historical information. This embodies the comprehensive spirit of the park, combining relaxation, fitness and cultural entertainment (Fig.01).

The location of the park's equipment is based on the terrain and different users' needs. Exercise equipment for seniors is located in lower areas, while the children's equipment is located closer to the entrance.

The circulation system is divided into three levels for the convenient access of vehicles and pedestrians. The plant species selection follows the principle of native planting, thus park users can experience seasonal change from the foliage.

By respecting the existing condition of the site and working with the distinctive mountainous terrain, the design provides a comprehensive city park for the public to enjoy (Fig.02-08).

查尔斯·沙（校订）
English reviewed by Charles Sands

图 04 次入口景观效果图
Fig 04 Main Entrance Perspective

图 05 景观节点效果图一
Fig 05 Vision of Landscape Node-1

图 06 景观节点效果图二
Fig 06 Vision of Landscape Node-2

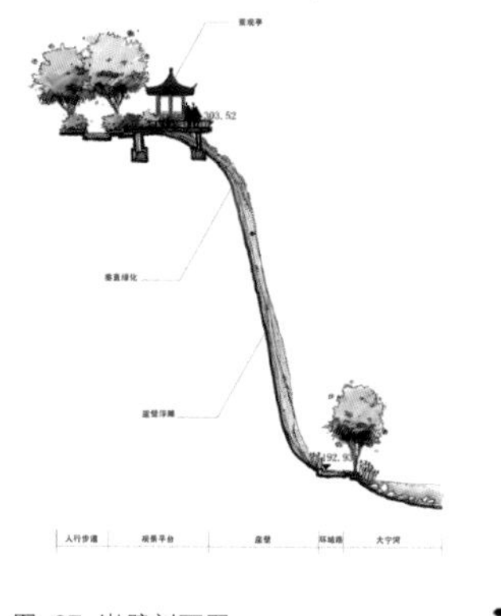

图 07 崖壁剖面图
Fig 07 Cliff Section

318.45

图 08 主入口断面图
Fig 08 Main Entrance Section

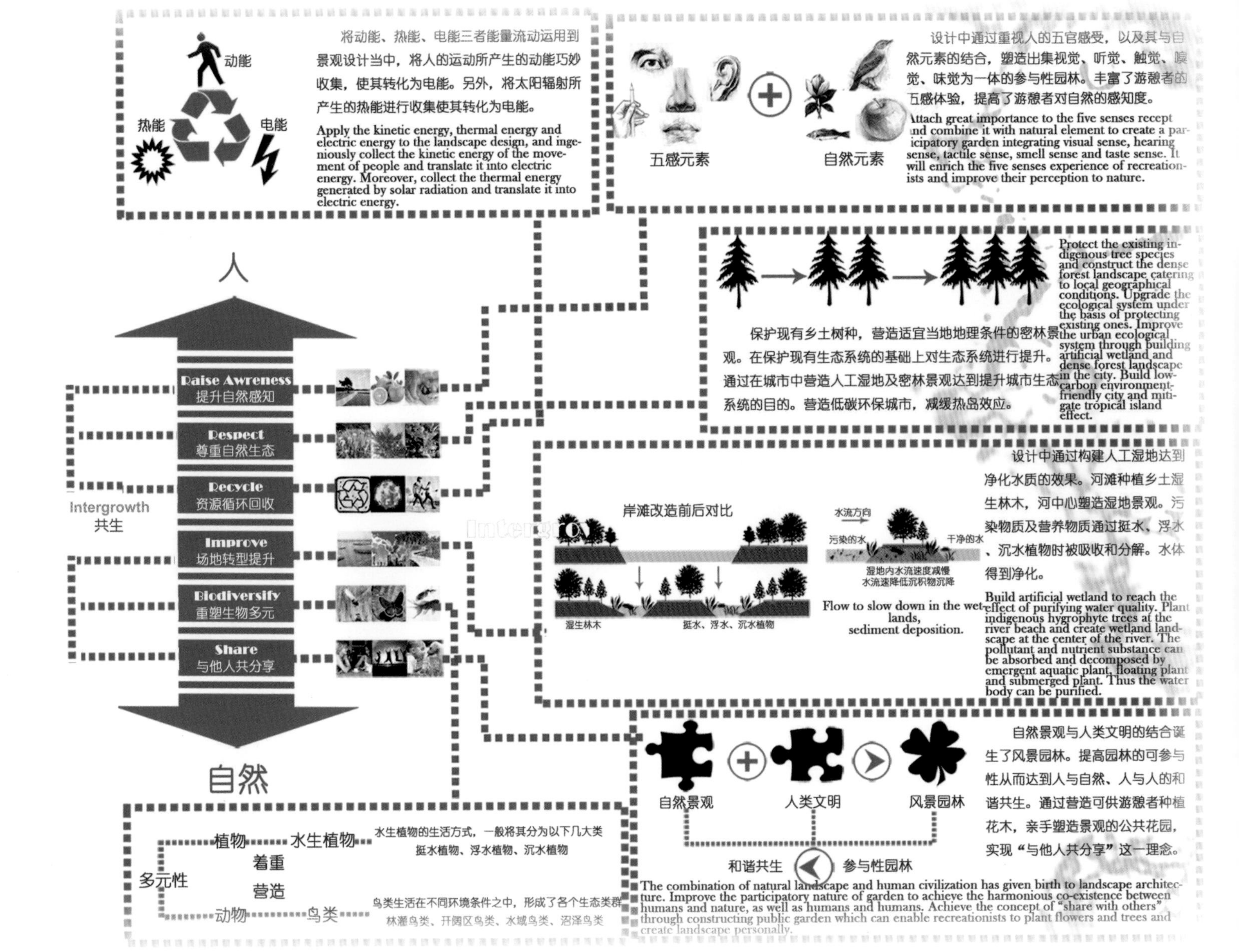

图 01 概念构思
Fig 01 Conception

INTERGROWTH
——广安西溪峡谷生态体验公园景观设计

INTERGROWTH
——THE XIXI CANYON ECO-EXPERIENCE PARK, GUANG'AN

中文标题：Intergrowth——广安西溪峡谷生态体验公园景观设计
组　　别：本科设计类
作　　者：陈心怡 陈英轩 张长亮
指导教师：郭　麗 刘维东
学　　校：四川农业大学
学科专业：园林
研究方向：风景园林
分　　类：2012“园冶杯”国际竞赛本科设计类作品二等奖

Title: Intergrowth — The Xixi Canyon eco-experience park, Guang'an
Degree: Design Group
Author:Xinyi Chen, Yingxuan Chen, Changliang Zhang
Instructor: Lu Guo, Weidong Liu
University: Sichuan Agricultural University
Specialized Subject: Landscape
Research: Landscape
Category: The Second Prize in Design Group of 2012 "YuanYe Award" International Architecture Graduation Project/Thesis Competition

图 02 平面图
Fig 02 Planform

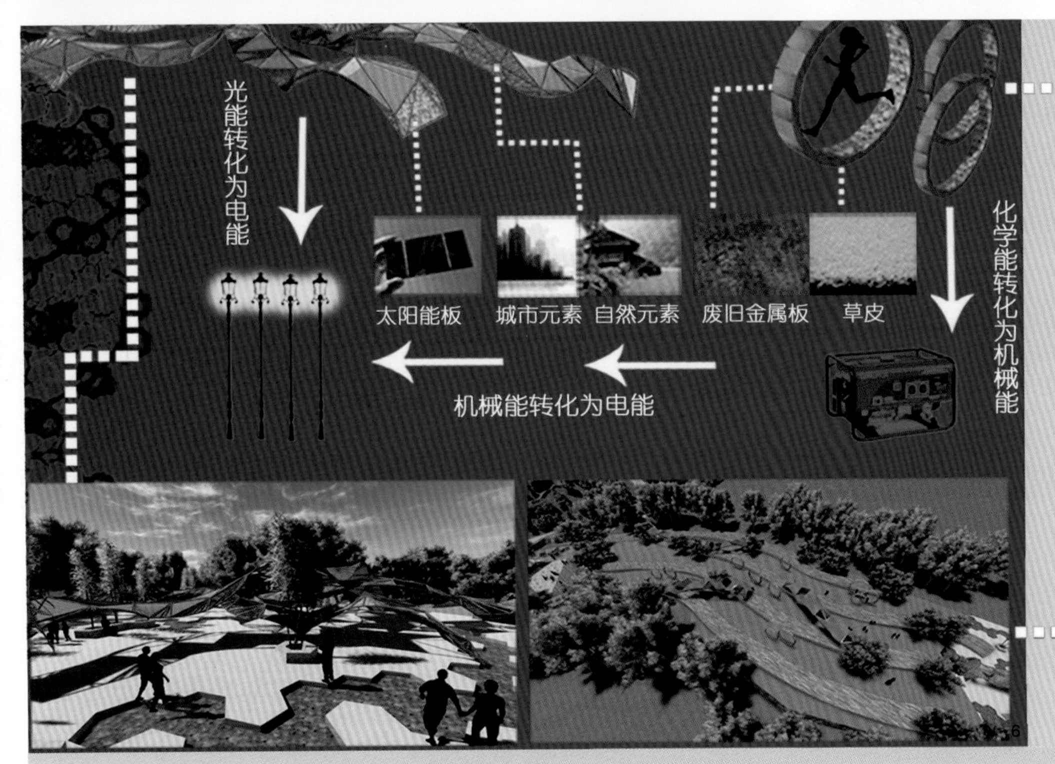

图 03 参与性花园效果图
Fig 03 Participatory garden rendering

图 04 步行环效果图
Fig 04 Walk ring rendering

图 05 循环广场鸟瞰图
Fig 05 Circulatory recycling square bird's eye view

图 06 循环广场分析图
Fig 06 Circulatory recycling square analysis chart

图 07 五感体验区鸟瞰图
Fig 07 Five-Sense Experience Area

图 08 生态 DNA 鸟瞰图
Fig 08 Ecological DNA bird's eye view drawing

图 09 人工湿地区
Fig 09 Artificial Wetland Area

图 10 鸟岛天堂效果图
Fig 10 Bird Paradise Ecological Island rendering

作品简介：

本设计为四川广安西溪峡谷生态体验公园景观设计。设计场地位于广安城区偏东，占地 14hm^2。随着城市的发展，人与自然的冲突日渐增多，环境问题日益严重。认知，理解自然，实现人与自然的共生是本设计的设计目的．通过“Intergrowth 共生”这一主题，引导出“提升自然感知”、“尊重自然生态”、“资源循环回收”、“场地转型提升”、“重塑生物多元”以及“与他人共分享”这 6 大理念，营造生态多元景观（图 01）。本设计的景观结构为一带两点（图 02）。一带为沿河景观带，两点为参与性公共花园（图 03–04）及循环回收广场（图 05–06）。沿河景观带沿岸及河中共分布有 5 大景观点，分别是：五感体验区（图 07）、生态 DNA（图 08）、人工湿地区（图 09）、鸟天堂生态岛（图 10）、密林景观区。景观形态特点以及结构的设计通过细胞以及 DNA 演绎而来，体现尊重生命，尊重自然这一思想。

Introduction:

Covering 140,000 square meters, the eco-experience park at Sichuan Guang'an Xixi Canyon is located on the east side of the Guangan urban area. Due to urbanization, conflicts between humans and nature have resulted in serious environmental problems. The design intention is to understand the natural symbiosis of man and nature. Transformation occurs through the theme of "intergrowth coexistence, "The design follows the principles of "enhancing the natural", "respecting ecology", "recycling and reusing resources"," enhancing the site"," rebuilding biological diversity " and "sharing among others". These 6 important ideas help to create a diverse ecological landscape (Fig. 01). The landscape structure is comprised of one zone and two focal points (Fig.02). The zone is the riverside landscape belt, while the two focal points refer to the participatory public garden (Fig.03-04) and the recycling square (Fig.05-06). The Five-Sense Experience Area is comprised of Five scenic spots distributed along the riverside landscape belt and within the river itself (Fig.07). The five sites are: the Ecological DNA area (Fig.08), the Artificial Wetland Area (Fig.09), Bird Paradise Ecological Island (Fig.10) and the Forest Landscape Area. The landscape morphological characteristics and structural design reflect the physical characteristics of cells and DNA. The park is ultimately about respecting life and nature.

查尔斯·沙（校订）
English reviewed by Charles Sands

5
7
8
9
10

“虽由人作，宛自天开”是中国古代造园大师计成在《园冶》中的名句，道出了造园的最高境界。以质量立身，以景效立命的天开集团恪守这份造园理念，抱着“为有限的城市空间营造无限的自然理想”的初衷，励志将公司打造成为“人居环境景观营造”的标杆式企业。天开园林首家公司由现任总裁陈友祥和副总裁谭勇于2003年在重庆联手创立，从此开启了天开时代的奋斗历程。2004年北京天开公司成立，天开集团进一步加快前进步伐，业务量逐年稳步增长，公司知名度也在大幅提升。2012年，天开完成了全国布局和业务升级，分公司覆盖中华各大区，北京、天津、上海、重庆、成都、长沙、哈尔滨等地。源于对景观工程的质量和景效的高要求，天开在行业内获得众多如万科、龙湖、纳帕、中旭、泰禾等高端合作伙伴的信赖，2012年3月，蓝光和骏地产也正式与天开签约结为战略联盟。时至今日，天开已发展成为一家集园林景观设计、园林工程施工、家庭园艺营造、苗木资源供应及石材加工为一体的领军式园林公司。我们的实力、品质和战略布局，能为客户有效降低沟通管理成本，更是全国地产企业所有项目卓越品质的保障。

China has a long history in Gardens and Landscape. “Materpiece of Nature, although Artificial Gardens”, as the first and the most acknowledged philosophy about garden building in the world, was initiated from the book Yuanye which was written by Ancient Chinese gardening master Jicheng, and Tiankai was named after it. Tiankai company was established in Chongqing in 2003 by President Chen youxiang and Vice-president Tan yong. We come from nature and back to natue, that’s the reason why we love nature gardens. Tiankai hopes to create infinite natural idea for the limited city. Today, Tiankai has become to comprehensive group corporation including garden construction, design, plants maintenance, seedling resources and so on. Tiankai has the second class qualification of garden construction, and B class qualification of landscape design. There are many international and outstanding designers in Tiankai. Tiankai is the best company in China in garden design and construction field. Tiankai has established 10 branches respectively in Beijing, Shanghai, Tianjin, Chongqing, Chengdu, Changsha and Haerbin. Tiankai's contruction business covers all over the country. The partners of Tiankai include Vanke, LongFor, Napa and Blue Ray, which are all very famous enterprises in China. We can effectively save communication and management cost for our clients and keep all the projects in consistent and excellent quality.

www.tkjg.com　tiankai@tkjg.com

华中农业大学园林植物遗传育种团队部分草花育成品种介绍

INTRODUCTION OF PARTIAL NEW CULTIVARS OF BEDDING FLOWERS BRED BY THE GROUP OF GENETIC IMPROVEMENT AND BIOTECHNOLOGY OF LANDSCAPE PLANTS IN HUAZHONG AGRICULTURAL UNIVERSITY

何燕红 傅小鹏 胡惠蓉 叶要妹 刘国锋 Yanhong He Xiaopeng Fu Huirong Hu Yaomei Ye Guofeng Liu

图 01 包满珠教授
Fig 01 Professor Manzhu Bao

团队带头人 – 包满珠教授

Leader of the group -- Professor Manzhu Bao

包满珠，男，1963 年生于甘肃省漳县。1980–1984 就读于北京林业大学园林专业，获园林学士学位；1991 年于北京林业大学获园林植物博士学位，导师陈俊愉教授。1994–1995 由国家公派赴英国德蒙福特大学 (De Montfort University) 进行了为期一年的博士后研究。1998–1999, 在荷兰 Vrije University of Amsterdam 做高级访问学者 6 个月。2002 年赴美国俄勒冈州立大学 (Oregon State University) 进行为期三个月的访问研究。1984 年进入华中农业大学工作，现为华中农业大学园艺林学学院院长。

自参加工作以来，一直工作在园林教学科研第一线，主要从事园林植物遗传育种及生物技术领域的研究，取得了显著成绩，同时在城市生物多样性、人居环境等方面也有一定建树。在他的带领下，华中农业大学园林植物遗传育种研究团队已成为在国际上有较大影响的创新团队。

包满珠教授先后主持国家自然科学基金、国际科学基金、国家 863 计划、教育部新世纪优秀人才计划、教育部重点项目及农业部 948 项目等多项。其研究成果在国内外学术刊物上发表论文 160 多篇；获得国家发明专利 8 项，审定了一系列草花新品种。

包满珠教授现任中国园艺学会副理事长，湖北省观赏园艺学会理事长，《园艺学报》副主编。

Professor Manzhu Bao, born in 1963 in Zhangxian,Gansu Province , received Ph.D from Department of Landscape Plants , Beijing Forestry University in 1991, mentored by Professor Junyu Chen. He did his postdoctoral training at De Montfort University in 1994–1995.As a senior visiting scholar, he studied for six months at the Vrije University of Amsterdam in the Netherlands during1998–1999 and three months at Oregon State University in USA in 2002. He joined the Huazhong Agricultural University in 1984, and now he is the dean of the College of Horticulture and Forestry Sciences.

Professor Bao has been committed to teaching and research work of Landscape Plants, mainly engaged in the Biotechnology and Genetic Breeding of Landscape Plants, and has made remarkable achievements, he also researched in the urban biodiversity, human settlements environment and other aspects.

Professor Manzhu Bao has successively presided over many research projects funded by National Natural Science Foundation of China , International Foundation for Science, the national 863 plan, the ministry of education in the new century of talents scheme, the ministry of education and the ministry of agriculture etc. He has published over 160 articles in domestic and international academic journals , eight national invention patents have been approved, a series of new cultivars of bedding flowers have been commercialized.

Prof. Bao is vice chairman of Chinese Society for Horticultural Science, chairman of Hubei Society for Horticultural Science, associate editor–in–chief of Acta Horticulturae Sinica.

图 02 草花新品种现场审定
Fig 02 Field evaluation of the new cultivars of flowers

图 03 品种审定证书
Fig 03 Cultivar certificates

图 04 华中农业大学园林植物遗传育种研究团队
Fig 04 The group members

华中农业大学园林植物遗传育种团队

The group of Genetic Improvement and Biotechnology of Landscape Plants

现有固定研究人员 11 人，教授 3 人，副教授 4 人，高级工程师 1 人，讲师 3 人，具博士学位的 10 人，年龄构成以 30–50 岁的教师为主，形成了人员结构和研究方向布局完善，年龄结构和职称结构合理，学位层次高，有强势发展潜力的研究集体。此外，还有近 60 名的博士、硕士研究生在实验室从事相关科研工作。

针对我国目前城市美化的重要草花品种产权主要由国外控制，大多花卉种子依赖进口的局面，华中农业大学园林植物遗传育种研究团队在包满珠教授的指导下，对矮牵牛、孔雀草、万寿菊、百日草、三色堇、羽衣甘蓝、石竹属等重要花卉的资源进行了系统收集，收集国内外资源 150 余份，并初步确定了各草花 F1 或品系在武汉地区的主要观赏性和适应性。通过系谱法获得了系列优良自交系共计 100 余份、获得 44 个雄性不育系。建立了三色堇、矮牵牛、孔雀草、羽衣甘蓝、石竹等的再生体系。进行矮牵牛、三色堇、石竹、百日草、孔雀草的杂交优势育种、远缘杂交育种，选出了百余个优良的杂交组合。现已审（认）定的草花品种有 10 个，其他系列品种目前正在进行品比，即将进入审定阶段。

The group has a faculty of 11 permanent researchers, aged between 30-50 years old, including 3 professors, 4 associate professors, 2 senior engineers and 3 Lecturers.10 of them have a doctoral degree, we can concluded that it is a research group with a strong potential for development. In addition, there are nearly 60 doctor and master students engaged in related research work in this group.

Nowadays, most of the flower cultivars, applied in the urban greening of China, are foreign cultivars, and flower seeds depend on imports. To resolve this situation, the group studied on developing new cultivars under the guidance of professor Bao. Through years of effort, more than 150 domestic and foreign accessions of *Petunia hybrid, Tatetes patula, Tagetes erecta, Zinnia elegans, Viola ×wittrockiana, Brassica oleracea* var. *acephala* f. *tricolor, Dianthus chinensis* are collected and their main ornamental characteristics and adaptability are initially identified in Wuhan. More than 100 inbred lines and 44 male sterile lines were obtained by pedigree method. Regeneration systems of these plants are established. Sexual cross breeding and wide cross breeding have been carried out and more than 100 excellent hybrid combinations have been selected. Until 2012, 10 cultivars have passed the certification of varieties, and is promoting to the market.

矮牵牛
Petunia hybrida

图 01 ‘红星闪闪’的花（上）、单株（中）、群体（下）
Fig 01 The flower (up), individual (middle) and population (down) of ‘Hongxing Shanshan’

矮牵牛（*Petunia hybrida*）为茄科矮牵牛属植物，主要由原产南美的膨大矮牵牛（*P. integrifolia*）和腋花矮牵牛（*P. axillaris*）杂交而来，现世界各地广泛栽培。矮牵牛花冠喇叭形，花色极为丰富。露地应用，常作一年生栽培，花期 4–10 月；温室栽培可多年生长，且全年开花。

品种介绍

1、矮牵牛精灵系列‘红星闪闪’

春播后 60 天左右进入始花期，80 天左右呈现盛花状态。茎秆直立，株型紧凑，分枝性强。盛花期株高 20–25 cm，花冠漏斗状，花径 5.5–6.5 cm，鲜红色，花梗稍短。花期长达 3 个月以上，且花开不断。精灵系列‘红星闪闪’品种颜色鲜艳，典型的中国红色，加之花瓣 5 裂，像一枚闪亮的红星，象征着我们繁荣富强、欣欣向荣的祖国，寓意深刻，令人鼓舞。详见图 01。

2、矮牵牛美誉系列‘佳人’

春播后约 60 天开花，盛花时株高 20–26 cm，平均冠幅 30 cm 左右；茎秆直立，株型紧凑，分枝数多达 12–16；花冠漏斗状，花径 6–7 cm，玫红色。花期长，自然栽培 4–10 月均可开花，温室栽培可周年开花。详见图 02。

生态习性

喜温暖和阳光充足的环境。不耐霜冻，生长适温 13–18 ℃，冬季温度低于 4 ℃，植株生长停止，夏季能耐 35 ℃以上的高温。怕雨涝，花期雨水多，花朵易褪色或腐烂；不耐积水，盆栽矮牵牛宜用疏松肥沃和排水良好的砂壤土。

园林应用

矮牵牛的花语是安全感，有你就很温馨，代表着人与人之间一种美好、温馨的感情，象征着和谐、美满的人际关系。矮牵牛在全世界有广泛栽培应用，适宜于春、夏、秋季花坛（图 03），被誉为“花坛花卉之王”，还可运用于花境及建筑物边缘点缀、悬挂装饰、花槽种植、庭院布置、室内装饰、集中摆放（图 04）及自然式布置。

繁殖技术要点

矮牵牛主要采用播种繁殖，但一些重瓣品种需进行无性繁殖，如扦插和组织培养。

播种繁殖：矮牵牛常作一年生栽培。播种时间视上市时间而定，如 5 月需花，应在 1 月温室播种；10 月用花，需在 7 月播种。时间还应根据品种不同进行适当调整。矮牵牛种子细小，千粒重在 0.1 g 左右，发芽适温为 20–25 ℃。穴盘播种，播种土为 3 份泥炭土加 1 份珍珠岩，之后覆盖一薄层细土或者不覆土，细孔喷雾器喷水，放在有漫射光的室内，约 10 天左右发芽，当出现真叶时，注意保湿通风。

栽培技术要点

矮牵牛在早春和夏季需充分灌水，但又忌高温、高湿。土壤肥力应适当，土壤过肥，则易于旺盛生长，以致枝条伸长倒伏。具体栽培要点如下：

(1) 定植。要求 pH 6.0–7.5 土壤，不宜多肥，以防疯长倒伏。小苗长 4–5 片叶时就可定植。盆土以园土、泥炭、粗糠灰按 1:1:1 配制。栽后立即浇一次透水。

(2) 整枝摘心。待其长至 7–9 叶时，需摘心一次。为多开花或整形需要可剪去植株上正在孕花上长的延长枝。盛花期后及时将枝条剪短，剪后 4–6 天，其芽能萌发。第一次留基部 2–3 节，以后视枝条疏密情况，再生枝上密留 1 节、疏留 2–3 节。

(3) 浇水。矮牵牛耐旱怕涝，需见干见湿，浇就浇透。小花型较耐雨水，大花型不耐雨水，注意控制。

(4) 施肥。根据不同生育期给以补肥，特别是剪枝后必须结合浇水施以腐熟的人粪尿或饼肥水。平时薄肥多施，生长旺期每隔半月施 1 次。

(5) 病虫防治。常见病害有：①白霉病，及时摘除病叶，喷施 75% 百菌清 600–800 倍液。②青枯病，为细菌性病害。应避免连作，用土须严格消毒，室内栽培需经常通风。③茎腐病，注意控制温度和浇水，加强通风；及时销毁病株、病叶。④花叶病，为病毒性病害，发现病株及时烧毁，种子和工具可用 10% 漂白粉消毒 20 分钟。虫害主要有蚜虫、蜗牛、蛞蝓等。蚜虫可用 40% 乐果 1000–1500 倍液，或 1500–2000 倍杀虫菊酯喷施。蜗牛和蛞蝓应及时施用蜗克灵防治。

图 02 ‘佳人’的花（上）、单株（中）、群体（下）
Fig 02 The flower (up), individual (middle) and population (down) of ‘Jiaren’

The garden petunia *(Petunia hybrida)*, an ornamental hybrid of the family Solanaceae widely cultivated around the world, is derived from *P. integrifolia* and *P. axillaris*, two Petunia species endemic to South America. The flowers of petunia are trumpet–shaped, with many different colors. Petunia is a perennial species but usually cultivated as an annual plant when cultivated outdoor, especially in cool region. The flowering period of petunia is during April to October, but if cultivated in greenhouse they will bloom all year round.

Introduction of cultivars

1.The Fairy Series 'Hongxing Shanshan' ('Red Shining Stars')

The cultivar began to flower about 60 d after sowing in spring, and will get into the full blooming stage after 20 more days. It has upright stems and strong capability of branching with compact plant form especially when pinching is applied. The plant height is about 20 to 25 cm. The flowers are trumpet–shaped and 5.5 to 6.5 cm in diameter with brilliant red color. The flowering period can last for over three months, with continuous flower formation. The vivid color of the 'Hongxing Shanshan' represents the classic Chinese red and the five limbs of the petal resemble a shining star, which symbolize a profound message that our motherland is rich, strong and prosperous. Please see（Fig.01）.

2.The Good Reputation Series 'Jiaren' ('Beauty')

The cultivar is going to bloom about 60 d after sowing. At the full blooming stage, the plant height is about 20 to 26 cm with the plant width about 30 cm. They have upright stems and possess a strong ability of branching with the number of branches up to 12 to 16, which lead to a compact plant form. The flowers are funnel–shaped with a rose color and about 6 to 7 cm in diameter. The flowering period can last from April to October if cultivated in natural conditions, and in the greenhouse, they will bloom all year round. Please see（Fig .02）.

Ecological habits

Petunias like warm and sunny conditions, while not resistant to frost. The most suitable temperature for their growing is between 13 to 18 ℃ . If the temperature is below 4 ℃ in winter, the plants will stop to grow. However, the plants can tolerate high temperature over 35 ℃ . They are afraid of rain and waterlogging. During their flowering period, the flowers are going to fade and rot if there are a lot of rains. It is better to use fertile and sandy soil of well drainage when cultivate petunias in pots.

Landscape applications

The flower language of petunia is safety and warmth which represents a friendly and loving feeling among people and signifies a harmonious and happy interpersonal relationship. Petunias are very popular all over the world and very suitable to be planted in

图 03 矮牵牛盛花花坛，摄于浙江虹越
Fig 03 Clustering flower bed of petunia, photographed in Zhejiang Hongyue Compan

flower beds in spring, summer and autumn (Fig.03). Due to their wide applications, they have a good reputation as "the king of bed flower". Besides applied in flower beds, petunia can also be planted as the edge decorations in flower borders or around the buildings, used for hanging baskets or planters, planted for yard, square or indoor decorations (Fig.04), or arranged naturally.

Propagation

Sexual propagation (sowing) is the main method of propagating petunias, but vegetative propagation such as cutting or tissue culture are the proper ways for some special cultivars like sterile double–flower ones.

Petunias are usually cultivated as annual plants. The sowing time is determined by the time they are needed. For example, if they are needed in May, then they should be planted in January in greenhouse, but if they are going to be on the market in October, then they should be sowed in July. In addition, the sowing time should also be adapted according to the characters of different cultivars. The seeds are very small and light, thus 1000–grain weight is only about 0.1 g. The most suitable temperature for germinating is between 20 to 25℃ . Use plugs to sow, the soil should be three quarters of peat and one quarter of perlite. After sowing, a thin layer of medium can be used to cover the seeds but this is not very necessary. After watering with a sprayer, the plugs are put in greenhouse with sunshine. About 10 d later, the seeds will germinate, then the true leaves come out. During this stage, the humidity of medium should be proper and environment should be ventilated.

Cultivation

In early spring and summer, it is important to give enough water to the plant, but avoid of high temperature and high humidity. The fertility of soil needs to be proper. If the soil is too fertile, the plant will grow very quickly and become susceptible to lodging. The key points for cultivation are followed.

1)Field planting. The pH value of soil should be 6.0–7.5. Never to be too fertile. Field planting can be carried out when the seedlings have 4–5 true leaves. The composition of soil is 1:1:1 of garden mould, peat soil and rice hull ash. Water immediately and thoroughly after planting.

2)Pruning and pinching. Pinching after the plant has 7–9 leaves. In order to make more flowers or keep a good plant form, the elongated long shoots with flower buds need be cut down. After the full flowering period, cut the shoots short in time. Four to six days later, the bud will sprout. The first cutting should keep 2 or 3 nodes. Later, the cutting should depend on the density of branches. If the axillary shoots are very dense, then just keep one node, but if the shoots are less, then keep 2 or 3 nodes.

3)Watering. Petunias are resistant to drought but afraid of waterlogging. Thus, water after the soil is dry, and once water, water thoroughly. The cultivars with small flowers are more resistant to rains, the vise versa.

4)Fertilizing. Supply different amounts of fertilizer according to different growing stages. Especially after cutting branches, it is necessary to water the plants with night soil or oil cake fertilizer. Usually, the right way should be fertilize more times with low concentration. In fast growing period fertilize half a month.

5)Disease and pest control. The common diseases are:

a)White mould. Remove all the sick leaves in time, and spraying 75% Chlorothalonil diluted 600 to 800 times.

b)Bacterial wilt. It's a disease caused by bacterial. Prevent continuous cropping, and sterilize the soil.

c)Stem rot. Note to control temperature and watering, and to strengthen ventilation. Destroy the diseased plants and leaves in time.

d)Mosaic diseases. It's caused by virus. Whenever find the disease, burn the ill plants immediately. All the seeds and tools should be disinfected with 10% bleaching powder for 20 minutes.

e)Insects. Aphid, snail and slug are the main destructive insects. In order to kill aphids, spray 40% Rogor diluted 1000 to 1500 times or spray Insecticidal Pyrethroids diluted 1500 to 2000 times. For snail and slug, Wokeling can be used.

图 04 广场盆栽观赏，摄于华中农业大学
Fig 04 Pot cultivating on square,photographed in Huazhong Agricultural University,Wuhan

图 01 三色堇繁星系列顶端叶片
Fig 01 The upside leaves of pansy Fanxing series

图 02 三色堇繁星系列中部叶片
Fig 02 The middle leaves of pansy Fanxing series

图 03 ‘猫咪’花朵
Fig 03 Flower of 'Maomi'

图 04 ‘猫咪’植株
Fig 04 Plant of ' Maomi'

三色堇
Pansy

三色堇，*Viola* × *wittrockiana*，堇菜科堇菜属，又名蝴蝶花、猫儿脸、鬼脸花。原产欧洲，现世界各地广泛有栽培，为园林中常见的二年生花卉，花期为 3 月至 6 月。

品种介绍

大花三色堇繁星系列品种具有共同的杂交母本，故其种苗及株型特征较为一致：紧凑，分枝量大。茎绿色光滑，直立生长。基生叶近心形边缘浅波状，茎生叶较狭长，托叶大，抱茎而生，基部成羽状裂，叶长 4–5 cm，宽 2–2.5 cm。

各品种开花差异明显，特征见（图 01–02）。

1. 繁星系列‘猫咪’

‘猫咪’（*Viola* × *wittrockiana* 'Maomi'）株高 22 cm，花黄色，上部两瓣纯黄，侧瓣及唇瓣黄色有深色条纹，花腋生，花梗长 8.5 –10 cm，花直径 3.5 –5 cm；花期 145 –150 天且花量很大；千粒重为 0.95 g，寿命 3 年。播后 120 天即可开花。详见（图 03–05）。

2. 繁星系列‘微笑’

‘微笑’（*Viola* × *wittrockiana* 'Weixiao'）株高 20 cm，花紫色，上部两瓣深紫，侧瓣及唇瓣紫色有白色斑块内有深色斑纹，花腋生，花梗长 6–8 cm，花直径 4–5 cm；花期 160–165 天且花量很大；千粒重为 0.85 g，寿命 3 年。播后 110 天即可开花。详见（图 06–08）。

生态习性

性喜冷凉，较耐寒，不耐酷暑：生长适温为 15–25 ℃，当气温连续高于 30 ℃时，长势明显减弱，且开花稀少，开花不良。怕水湿，忌干旱：夏季多雨或土壤排水不良时，植株细长，花朵变小；过于干旱，也会对花朵的数量、花色、花径大小等产生直接影响。耐瘠薄，忌连作：在土层深厚的沃土中生长最佳，过于瘠薄会影响开花效果，应适时轮作。

观赏应用

三色堇株型低矮、叶丛繁密、花型独特、花色鲜艳、斑纹俏丽、花期较长，是冬春园林的重要花材，被誉为“花坛皇后”。除在花坛（图 09）、花台、花丛、花境中露地应用外，也可用于盆栽、切花（如“襟花”）、压花等室内装饰及艺术形式。

繁殖技术要点

三色堇主要采用秋季播种繁殖（图 10），也可进行春播或扦插繁殖。

秋播于 8 月下旬 – 9 月中旬进行，选择排水良好、疏松的播种基质，播后需严密覆盖 1 cm 左右基质，以不见种子为度，并置于避光处；浇透水，注意保湿；发芽适温为 15–20 ℃，一般 7 天左右即可发芽出土，发芽率 70% 以上。

栽培技术要点

1 **移栽定植**：三色堇幼苗生长极快，容易徒长，待幼苗长至 2 对真叶时要及时移栽定植，移栽前 2–3 天将种苗置于室外进行炼苗，移植的操作宜在傍晚或阴天进行，以提高成活率。

2 **光照条件**：三色堇为阳性植物，生长和开花均要求阳光充足。

3 **温度控制**：三色堇生长适温为 15–25 ℃，不耐酷暑。

4 **水肥管理**：浇水要坚持“见干见湿”，一般上午浇水较下午浇水好。营养生长期 7–10 天施一次液肥，现蕾后停止施氮肥，开花期间可酌情施入磷酸二氢钾等复合肥。去残要及时，需从花葶基部摘除枯萎的花朵，以减少养分消耗；去残后要勤浇水，并且追肥 2–3 次，可以诱发新蕾，延长观赏期。

5 **病害防治**：三色堇常见的病害有苗期的猝倒病，生长期的灰霉病、茎腐病、叶斑病；虫害主要有：蛞蝓、蚜虫、菜青虫等。需要在播种前做好种子消毒和土壤消毒，在病虫害发生初期及时酌情进行药物防治。

Pansy (*Viola × wittrockiana*), a member of the viola family (Violaceae), is called hudiehua (means butterfly flower), maoerlian (means cat face) and guilianhua (means grimace flower) in China. Pansies are the cultivated form of viola native to Europe and are widely planted all over the world. Blooming from March to June, pansies are used extensively in landscape as biennials.

Introduction of cultivars

Bred by crossing from the same female parents, the traits of seedlings and plant forms within Fanxing series of pansies are consistent: plants are compact and very much branched; stems are green, smooth and erect; basal leaves are round and nearly cordate while stem leaves are lanceolate; large stipules are pinnately parted towards the base; foliage is about 4–5 cm tall and 2–2.5 cm wide. However differences can be found obviously among cultivars in their flowering traits with details illustrated as follows(Fig.01–02).

1.*Viola ×wittrockiana* Fanxing series 'Mao mi'

Viola × wittrockiana 'Maomi' is 22 cm tall with yellow flowers. Two superior petals are pure yellow, while two lateral petals and a inferior petal are yellow with dark stripes. Flowers are axillary and 3.5–5.0 cm in diameter with 8.5–10 cm long peduncles. Blooming peiod can last 145–150 d with profuse flowers. The seeds are 0.95 g of kilo–grain weightiness with 3–year life. First flowering begins 120 d after sowing (Fig.03–05).

2.*Viola × wittrockiana* Fanxing series 'Weixiao'

Viola × wittrockiana 'Weixiao' is 20 cm tall with purple flowers. Two superior petals are deep purple, while two lateral petals and a inferior petal are purple with white plaque having dark stripes. Flowers are axillary and 4–5 cm in diameter with 4–5 cm long peduncles. Blooming peiod can last 160–165 d with profuse flowers. The seeds are 0.85 g of kilo–grain weightiness with 3–year life. First flowering begins 110 d after sowing (Fig.06–08).

Ecological habits

Pansies prefer cool climate, can't tolerate hot summer but respond well to coldness. They perform best in temperature vary from 15 ℃ to 25 ℃ . Temperature above 30 ℃ lasting for a long time can lead to rapid growth decline and sharp decrease in fowering. Pansies are susceptible to both wet and drought. Plants tend to be slender and flowers small in rainy summer days or poor–drained soil. Harms on quantity, color and diameter of flowers will occur when it is too dry. Pansies can stand barren soil but not continuous cropping. Plants will grow well in rich soil while flowering value usually decreased in barren soil. It's highly suggested that crop rotation be routinely accomplished.

Ornamental Uses

Pansies are famous for their dwarf appearence, flourishing foliage, unique flower form, bright color, fantastic stripes and long florescence, which make them of great importance in winter and spring landscape, and thus named "Queen of flowering beds". Besides their outdoor uses such as flowering beds(Fig.09), clusers and borders, pansies can also decorate indoor spaces as cut flowers (e.g. corsage) or potted flowering plants.

Propagation

Pansies can be propagated by sowing seeds(Fig.10) in autumn as well as in spring or by cutting.

Generally, seeds are sowed at late August to mid September. Well–drained and loose humus soil is essencial, more over covering soil with black plastic can help germination. Water thoroughly and mulch to conserve the soil moist. Seeds will germinate in one week when temperature is maitained at 15–20 ℃ with germination rate above 70%.

图 05 '猫咪' 园林应用
Fig 05 Landscape use of ' Maomi'

图 06 '微笑' 花朵
Fig 06 Flower of 'Weixiao'

图 07 '微笑' 植株
Fig 07 Plant of 'Weixiao'

图 08 '微笑' 园林应用
Fig 08 Landscape use of 'Weixiao'

Cultivation

1. Transplanting: Seedlings grow very fast and tend to produce watershoots, so it's better to transplant early as the first set of true leaves appear. In order to improve transplanting survival rate, seedlings should be hardened off outdoor 2–3 d before , and a cloudy day or evening is usually selected.

2. Light requirement: Pansy is a kind of heliophilous ornamental plant, it needs full sunlight both during growth and blooming.

3. Temperature requirement: Pansy performs best in temperature between 15 to 25 ℃ and can not tolerate hot summer.

4. Water and Fertilizer: Appropriate watering, keep the medium moist. It is better watering in morning than afternoon. In vegetative growth period, liquid fertilizer should be added every 7–10 d. If flower buds arise, nitrogenous fertilizer can't be used continously. And applying compound fertilizer (e.g. potassium dihydrogen phosphate) is suitable at blooming time. Deadheading should be in time, and withered flowers should be removed clearly from the basal of peduncles to reduce nutrient consumption. Watering regularly and fertilizing 2–3 times afterwards will induce new flower buds, and thus lengthen the blooming season.

5. Diseases and Insects: Pansy is subjected to a variety of fungal diseases, including cataplexy at seedling stage and gray mold, crown rot, leaf spots during growth period. Common insect pests are slugs, aphids and pieris, etc. Sterilization should be done to both seeds and soil before seeding. And the earlier inspect pests controlled, the better.

图 10 三色堇繁星系列种苗
Fig 10 Seedlings of pansy Fanxing series

图 09 三色堇园林应用
Fig 09 Landscape uses of pansies

图 01 ‘晨韵橙色’的种子
Fig 01 Seed of ‘Chenyuncheng’
图 02 ‘晨韵橙色’的幼苗
Fig 02 Seedlings of ‘Chenyuncheng’
图 03 ‘晨韵橙色’营养生长期特性
Fig 03 The vegetative stage characteristics of ‘Chenyuncheng’

孔雀草
Tegetes patula

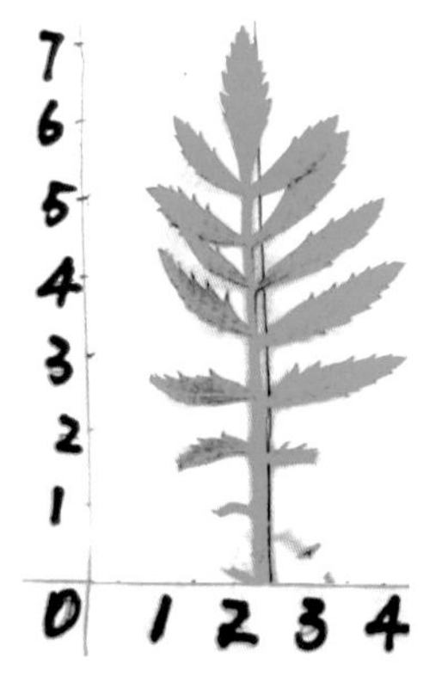

图 04 ‘晨韵橙色’的叶片
Fig 04 Leaf of ‘Chenyuncheng’

孔雀草，*Tegetes patula*，菊科万寿菊属，原产墨西哥，是一种适应性十分强的花卉。在我国很多地方也可见到，尤其南方更常见之。孔雀草是一年生草本花卉，花期从“五一”一直开到“十一”。

品种介绍

1、孔雀草晨韵系列‘晨韵橙色’

播后 40–50 天开花。盛花时株高 25–35 cm，冠幅 22–27 cm，茎杆粗壮直立，红色；株型紧凑，分枝数（三级分枝）较多，约 50–60；叶缘具油腺点，有异味；花序伸出叶平面，花蕾近似短圆柱状，绿色；头状花序顶生，莲状花型，舌状花 1 轮，橙色，基部带红晕，管状花橙色，先端 5 裂。花径 5.5–6.5 cm，花心径 2.0–3.0 cm，盛花期单株花朵数达 20–25 朵左右；观赏期可以持续 90–100 天；瘦果黑色、线形，顶部簇生灰白色的种毛；单果种子数逾 40 粒（武汉地区，自然结实率约 57%），千粒重约 3.0 g。抗性强，能耐早霜。详见（图 01–08）。

2、万寿菊 × 孔雀草 驼铃系列‘驼铃黄色’

驼铃系列‘驼铃黄色’品种是万寿菊和孔雀草种间杂交而来，播后均 40–50 天开花，尤其夏季花期比亲本提前 20 天左右。盛花时株高 20–30 cm，平均冠幅 30 cm 左右，茎杆粗壮直立，红色；分枝数（三级分枝）较多，约 45–55，株型紧凑；叶缘具油腺点，有异味；花蕾近似短圆柱状，绿色；头状花序顶生，莲座状花型，重瓣（3–4 轮舌状花），舌状花和管状花均为黄色，花径 4.5–5.5 cm，比父本增大约 1.0 cm，管状花径 1.0–1.2 cm；盛花期单株花朵数约 30–40 朵，盛花期长达到 120–130 d；不结实；抗热性强，夏季不易徒长，能耐早霜，抗寒性增强。详见（图 09–15）。

生态习性

一年生草本植物，喜温暖、阳光充足的环境，生长温度 10–38 ℃，最适温度为 15 –30 ℃。对土壤要求不严，但忌 pH 值小于 6 的酸性土。

应用

孔雀草花语是兴高采烈，爽朗活泼。孔雀草株型矮小紧凑，花量丰富，开花时间长，花色艳丽，适应性强，观赏性好，已逐步成为花坛、庭院的主要花卉之一。最宜作花坛、花丛、花境边缘材料等栽植，也可盆栽和作切花（图 16–17）。此外，孔雀草还富含生物活性物质，为重要的经济作物，其花瓣中的叶黄素具有抗氧化活性，根提取物中的酚类化合物（类黄酮和酚酸）和噻吩类化合物分别具有杀线虫活性和抑菌活性，广泛应用于农业生产。

繁殖技术要点

孔雀草用播种和扦插繁殖均可。

孔雀草的播种要求气温高于 15 ℃（或有加温保温条件），因此长江中下游一带的播种期一般在 2 月下旬 –3 月上旬和 7 月中下旬。气候暖和的南方可以一年四季播种，在北方则流行春播。播种宜采用疏松的人工介质，床播、箱播育苗，有条件的可采用穴盘育苗。经消毒处理，播种后稍稍覆盖，孔雀草的发芽适温 22–24 ℃。

扦插繁殖可于 6–8 月间剪取长约 10 cm 的嫩枝直接插于沙床，遮荫。夏秋扦插的 8–12 月开花。扦插不论插地或插床（盆）均可成活。

图 05 ‘晨韵橙色’头状花序俯视
Fig 05 The elevation of capitulum for ‘Chenyuncheng’

图 06 ‘晨韵橙色’头状花序侧视
Fig 06 The side elevation of capitulum for ‘Chenyuncheng’

图 07 ‘晨韵橙色’单株盛花期状态
Fig 07 The blooming stage of individual plant of ‘Chenyuncheng’

图 08 ‘晨韵橙色’群体效果
Fig 08 The group appearance of ‘Chenyuncheng’

栽培技术要点

孔雀草喜阳光，但在半荫处栽植也能开花。它对土壤要求不严。既耐移栽，又生长迅速，栽培管理又很容易。但要种植好孔雀草，仍要注意做到以下几点：

（1）摘心处理：带孔雀草第一和第二次现蕾时，均抹除花蕾，促进植株营养生长，枝繁叶茂。

（2）光照调节：孔雀草为阳性植物，要求阳光充足。

（3）温度控制：上盆后温度可降低至 18 ℃，经几周后可以降至 15 ℃，开花前后可低至 12–14 ℃，这样的温度对形成良好的株形是最理想的，但在实际生产中可能难以达到此条件。一般来讲，只要在 5 ℃以上就不会发生冻害，15–30 ℃间均可良好生长。

（4）水肥管理：水份管理的关键是采用排水良好的介质，保持介质的湿润虽然重要，但每次浇水前适当的干燥是必要的，当然不能使介质过干导致植株枯萎。对于完全用人工介质栽培的，可 7–10 d 交替施肥一次。如果是以普通土壤为介质的，则可以用复合肥在介质装盆前适量混合作基肥。肥力不足时，再追施肥料。

（5）病害防治：孔雀草常见的病害有褐斑病、白粉病等，属真菌性病害，应选择好地栽培，并注意排灌，清除病株，病叶，烧毁残枝，及时喷锈粉宁等杀菌药。虫害主要是红蜘蛛，可加强栽培管理，在虫害发生初期可用 20% 三氯杀螨醇乳油 500~600 倍进行喷药防治。此外，孔雀草幼苗期还应特别注意茎腐病、疫病、蜗牛等。

Tegetes patula, belonging to the marigold genus of Asteraceae family and native to Mexico , is a well known ornamental plant widespread all over the world for its beautiful flowers and strong adaptability to adverse environments. *Tagetes patula* is used extensively in China especially in the south area. *Tagetes patula* is an annual plant, which flowers can bloom from May to October.

Introduction of cultivars

1. *Tagetes patula* Chenyun series ‘Chenyuncheng’

Tagetes patula ‘Chenyuncheng’ will bloom after planting 40–50 d. Its height can be varied from 25 cm to 35 cm, and plant width is from 22 cm to 27 cm, with erect and red stout stems. Full planttype has 50–60 branch numbers. Leaf has peculiar smell because of oil glands at the edge of the leaf. Flower bud appears green short cylindrical. The inflorescence is composed by a wheel of orange ray florets with red spots at the bottom and many orange disk florets with 5 cracks at the top of the corolla. The diameter of inflorescence is 5.5–6.5 cm, and the diameter of center disk florets is 2.0–3.0 cm, Maximum flower numbers per day can reach to 20–25, and the ornamental period can last 90–100 d. Black achenes are linear and have gray hairs. There are about 300 seeds per gram, each inflorescence can produce more than 40 seeds (In Wuhan, natural seed rate is about 57%), *Tagetes patula* ‘Chenyuncheng’ has strong adaptability to adverse environments(Fig.01–08).

2. *Tagetes erecta* × *patula* Tuoling series ‘Tuolinghuang’

Tagetes erecta × *patula* ‘Tuolinghuang’ is a hybrid cultivar came from interspecific hybridization between *Tagetes erecta* and *Tagetes patula*, it has strong heterosis, effectively improve the ornamental character of *Tagetes patula*. It will bloom after planting 40–50 d, and is 20 days earlier than the parents in summer. The height can be varied from 20 cm to 30 cm, and plant width is from 22 cm to 27 cm, with erect and red stout stems. Full planttype has about 45–55 branchs. Leaf also has peculiar smell. Flower bud appears green short cylindrical and yellow inflorescence is composed by 3 wheels of the ray florets and many disk flowers. The diameter of inflorescence is 4.5–5.5 cm, which is larger than the paternal, and the diameter of center disk florets is 1.0–1.2 cm. Maximum flower numbers per day can reach to 30–40. Flower ornamental period can last 120–130 d because of no seed. It has strong resistance to heat and cold(Fig.09–15).

Ecological habits

Tagetes patula likes warm and sunny environment. The growth

图 09 ‘驼铃黄色’的种子
Fig 09 Seeds of ‘Tuolinghuang’ ’
图 10 ‘驼铃黄色’的幼苗
Fig10 10 Seedlings of ‘Tuolinghuang’
图 12 ‘驼铃黄色’营养生长期特性
Fig 12The vegetative stage characteristics of ‘Tuolinghuang’

图 11 ‘驼铃黄色’的叶片
Fig 11 Leaf of ‘Tuolinghuang’

temperature is 10 ℃-38 ℃ and the optimum temperature is 15 ℃-30 ℃. Clay, sand or loam soil is adopted except for acidic soil which pH value is less than 6.

Application

Tagetes patula is elated, bright and lively flower. It has good ornamental value, for its dwarf appearence, full planttypes, vary color, longer florescence, more flowers. It can be used in flower bed, flower cluster, flower border, cut flowers and potted plant(Fig.16-17). Besides, the flower petals of *Tagetes patula* can be effectively utilized to produce lutein ester used as natural antioxidants. In addition, the roots of *Tagetes patula* can secrete the phenolic and thiophene compounds, which are widely employed as insecticides, fungicides and nematicides.

Propagation

Sowing seeds and cutting propagation can be used to propagate *Tagetes patula*.

Sowing seeds: Temperature for sowing seeds requires higher than 15 ℃, so in wuhan, the sowing time is late February and late July. In the south, we can sow all over the year because of the warm climate, wherease, in the north, we can sow in spring. Loose artificial medium is used and the optimum temperature for germination is 22-24 ℃.

Cutting propagation: In June to August, stem sections about 10 cm long directly inserted into sand bed, and than covered with shade. They will be flowering in August and December.

Cultivation

Pinching processing: the pinching processes are happen at the appearance of the first and second flower buds resulting in the lushly grow and more flowers.

Light requirement: *Tagetes patula* grows in full sun, high light intensity increases flower development rate and flower number.

Temperature requirement: To control the plant height, the temperature can be reduced to 18 ℃ after planted in the pot. A few weeks later, temperature can be reduced to 15 ℃, and be reduced to 12-14 ℃ in the period of flowering. In general, *Tagetes patula* can grow in the condition more than 5 ℃, and 15 ℃ to 30 ℃ is better for its growing.

Water: Keep appropriate watering and the moist medium. But it is a common practice to use water stress to harden plants.

Fertilizer: An appropriate amount of fertilizer is added to the medium as basal fertilizer. Topdressing is necessary when the fertility is low.

图 13 ‘驼铃黄色’单株
Fig 13 the blooming stage of individual plant of ‘Tuolinghuang’

图 14 ‘驼铃黄色’的头状花序
Fig 14 The capitulum of ‘Tuolinghuang’

图 15 ‘驼铃黄色’的盛花期群体效果
Fig 15 The group appearance of ‘Tuolinghuang’

图 16 广场盆栽装饰，摄于武汉华中农业大学
Fig 16 *Tagetes patula* used as potted plants in HuaZhong Agricultural University

Insects and Diseases: Leaf spot disease and powdery mildew are fungal disease, good cultivation and Xiufenning medicine can be used to control the disease. Pest is mainly red spider. The good cultivation and 20% dicofol EC can be used to manage pests. In addition, special attention should also be paid to stem rot disease, Phytophthora disease, the snail during the seedling period.

图 17 盛花花坛，摄于浙江虹越
Fig 17 *Tagetes patula* used in full blooming flowering beds in HongYue, ZheJiang

百日草
Zinnia elegans Jacq

百日草（*Zinnia elegans*），菊科百日草属，原产南北美洲，以墨西哥为分布中心，因其花朵硕大，色彩丰富（图 01–12），花型别致（图 13–15），现为世界各地均有栽培。

品种介绍：

1、曙光系列‘玫瑰红’

第一花序株高 34–41 cm，株高 44–55 cm，株型紧凑。叶长 6–11 cm、宽 5.5–6 cm。头状花序，单生于枝端，花梗长 6–12 cm，花头高 2.5–3.5cm，花心直径 0.8–2.4 cm，花头直径 8–11 cm，纵切舌状花一般为 5 轮，盛花期温度较高时舌状花轮数增加；总苞钟状，舌状花花瓣倒卵形、先端钝圆、玫红色，筒状花边缘分裂、黄橙色；花期约 75 天，盛花期花头数 7–9 朵。与亲本相比，播后约 44 天即可开花，花色艳丽持久，花径较大，盛花期花头数较父本多 3–4 朵，且长势较强，抗虫性强。1 g 种子约有 150 粒，寿命 3 年（图 16–19）。

2、梦幻系列‘绯红’

第一花序株高 26–30 cm，株高 35–45 cm，株型中等。叶长 6–12 cm、宽 5.3–7 cm。头状花序，单生于枝端，花梗长 3.5–7.2 cm，花头高 2.4–3 cm，花心直径 1.6–2 cm，花头直径 7–11 cm，纵切舌状花一般为 3–4 轮，如果盛花期温度较高舌状花轮数增加；总苞钟状，舌状花花瓣倒卵形、先端钝圆、绯红色，管状花边缘分裂、黄橙色；花期长达 79 d 左右，盛花期花头数 6–8 朵。1g 种子约有 130 粒，寿命 3 年。与亲本相比，植株高度较母本矮，播后约 48 天即可开花，花色艳丽，花径较大，盛花期花头数较母本多 2–3 朵，且长势较强，抗虫性强（图 20–23)。

生态习性：

一年生直立性草花，生长势强，喜阳光充足的环境，也能耐半阴，怕水湿。适应性强，耐瘠薄，生长适温 20–25 ℃，不耐酷暑，忌连作。

园林应用：

开花早，花期长，是常见的花坛、花境材料。矮生种可盆栽，高杆品种适合做切花生产（图 24–26）。

繁殖技术要点：

百日草用播种和扦插繁殖均可。

百日草的播种要求气温高于 15 ℃，因此长江中下游一带的播种期一般在 2 月下旬 – 3 月上旬和 7 月中下旬，在北方则流行春播。播种宜采用疏松的人工介质，床播、箱播育苗，有条件的可采用穴盘育苗。经消毒处理，播种后稍稍覆盖，百日草的发芽适温 15–20 ℃。

扦插繁殖可于 6–7 月间剪取长约 10 cm 的嫩枝直接插于沙床，遮荫。夏秋扦插的 8–11 月开花。扦插不论插地或插床（盆）均可成活。

栽培技术要点：

百日草生长势强，适应性强，耐瘠薄。要种植好百日草，主要做到以下几点：

（1）摘心处理：定植一周缓苗期过后要适当摘心，留 3 对真叶，并视植株生长及分枝情况来决定是否进行再次摘心，以促发腋芽生长，使侧枝增多，枝茂花繁，株形饱满。

（2）光照调节：百日草为阳性植物，生长和开花均要求阳光充足。

（3）温度控制：一般来讲，上盆后只要在 5 ℃以上就不会发生冻害，18–30 ℃间均可良好生长。

（4）水肥管理：水分管理的关键是采用排水良好的介质，浇水采用“见干见湿”的原则。对于完全用人工介质栽培的，可 7–10 d 交替施肥一次。现蕾后停止施氮肥；开花期间可酌情施入磷酸二氢钾等磷钾肥，促使花多色艳，并可提高种子品质。如果是以普通土壤为介质的，则可以用复合肥在介质装盆前适量混合作基肥。肥力不足时，再追施肥料。

（5）病害防治：百日草常见的病虫害主要有苗期猝倒病、生长期青枯病、茎腐病、叶斑病以及夜蛾、红蜘蛛、菜青虫、蜗牛等。需要在播种前做好种子消毒和土壤消毒进行防治，在病虫害发生初期及时酌情进行药物防治。

图 01–12 / Fig01–12

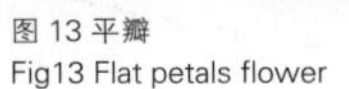

图 13 平瓣
Fig13 Flat petals flower

图 14 管瓣
Fig14 Tube-petal flower

图 15 丝瓣
Fig15 Filamentous flower

Zinnia elegans Jacq. (family Asteraceae, genus Zinnia) is native to North and South America, Mexico is the distribution center. It is famous for the chic, big and colorful flower(Fig 01–15); nowadays it is very popular with the whole world as a garden plant and cut flower.

Introduction of Cultivars

1. *Zinnia elegans* Shuguang series 'Meiguihong'

The individual is about 44–55 cm in height; its lanceolate leaves are about 6–11 cm long and 5.5–6.0 cm wide. They have attractive inflorescences (capitula) with about 5 whorls of bright rose florid florets which may increase with the increment of the temperature, the diameter of the capitula and center disk is about 8–11 cm and 0.8–2.4 cm respectively, the height of receptacle is about 2.5–3.5 cm, the length of pedicel is about 6–12 cm, what is more, the yellow tiny tubular florets is very fascinating. It will flowers 44 days later after sowing, its flowering period is from April to July when seeded in spring, but if seeded in summer it can also flower from August to November, so its flowering period is about 75 days, but the most amazing is that it has 7–9 capitulas in full-bloom stage. Compared with their parents, it has brighter colors, longer bloom periods, bigger and more capitulas, better growth and better abilities of insects-resistance. The thousand seed weight is 6.67 gram. The lifetime of seeds is about 3 years (Fig.03).

2. *Zinnia elegans* Dream series 'Feihong'

The individual of this cultivar is about 35–45 cm in height; its lanceolate leaves are about 6–12 cm long and 5.3–7.0 cm wide. They have attractive inflorescences (capitula) with about 3–4 whorls of bright rose florid florets which may increase with the increment of the temperature, the diameter of the capitula and center disk is about 7–11 cm and 1.6–2.0 cm respectively, the height of receptacle is about 2.2–3.0 cm, the length of pedicel is about 3.5–7.2 cm, what is more, the yellow tiny tubular florets is very fascinating. It will flowers 44 days later after sowing, its flowering period is from April to July when seeded in spring, but if seeded in summer it can also flower from August to November, so its flowering period is about 75 days, but the most amazing is that it has 7–9 capitulas in full-bloom stage. Compared with their parents, it has brighter colors, longer bloom periods, bigger and more capitulas, better growth and better abilities of insects-resistance. The thousand seed weight is 7.69 gram. The lifetime of seeds is about 3 years (Fig.03).

Ecological habit

Zinnia elegans is an erect herbaceous annual, it prefers sunny conditions but it can also survive from the semi-shade habitats, its optima growth temperature is from 20–25 ℃, it has the characteristics of infertile tolerant, but it cannot tolerant hot weather, flooded conditions and continuous cropping.

Application in garden

Zinnias are multifunctional plants, it is usually used in flower bed and border, however, the dwarf ones can be used as potted flower, and the long pedunculate ones can be used as cut flowers.

Breeding Techniques

In generally, *Zinnia elegans* can be bred by two methods: seeding and cutting.

The suitable temperature for germinating is 15–20 ℃ so in Yangtze river basin, it is usually seeded in late February or late July, but in the north of China the best time for seeding is spring. In general, we select loose soil with sterilization for sowing, seeds can be sown in seedbeds, the boxes or the polystyrene plugs, and then we scattered a thin layer of soil on the seeds, after 2–7 days, the seed will germinate out of the soil.

As for cutting propagation, we usually do it June or July, two points is very important: (1) select the 10 cm fresh shoot, (2) give some shading for shoot to protect them from losing too much water. After that, it will flower at August to November.

Cultivation

Even though *Zinnia elegans* is a vigorous and well-adapted plant, some points are still very important to make sure better growth.

(1) Pinching processing: Seven days after transplanting, we will do the first pinching processing, but three pairs of leaves should be kept; the second pinching depends on the growing-speed of the plant and the number of branches to make sure the plants grow lushly and have more flowers.

(2) Illumination regulation: *Zinnia elegans* is sun plant, which need enough light to satisfy growth and bloom.

图 18 曙光系列玫瑰红 F1 代单花头形态
Fig 18 morphological characteristics of single capitula of F1 hybrids

(3) temperature controlling: the phenomenon of freezing injury can happened when the temperature is lower than 5 ℃ , the plant can grow well in the condition of 18 –30 ℃ generally.

(4) Watering and fertilizing: too much water is avoided because this species don't like wet conditions; If the plant is cultivated with artificial medium, we can fertilize them seven to ten days one time, and stop fertilizing nitrogen at the presents of flower bud , at this special period, some quantitative monopotassium phosphate increase the number of branch, make the color brighter and improve the quality of the seed. If the medium consists of soil, we can mix a certain amount of compound fertilizer with the soil as base fertilizer before potting; the adding of fertilizer depends on the performance of the plant.

(5) Disease and pest control: One of the most dangerous threat to *Zinnia elegans* cultivars is from diseases, including seedling damping–off, bacterial wilt at growing stage, stem rot, leaf spot disease. Another threat comes from insects–attacking, including Noctuidae, Tetranychus cinnbarinus, Pierisrapae Linne, Fruticicolidae. In order to avoid it, disinfection to seeds and soil before sowing and pesticide should be taken into consideration in the initial stage when diseases and pests happen.

图 16 曙光系列玫瑰红母本花头形态
Fig 16 Morphological characteristics of capitula of female parent

图 17 曙光系列玫瑰红父本花头形态
Fig 17 morphological characteristics of capitula of male parent

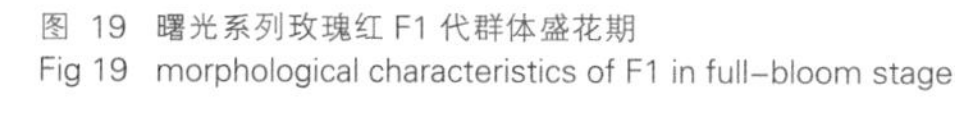

图 19 曙光系列玫瑰红 F1 代群体盛花期
Fig 19 morphological characteristics of F1 in full–bloom stage

图 20 梦幻系列绯红母本花头形态
Fig 20 Morphological characteristics of capitula of female parent

图 21 梦幻系列绯红父本花头形态
Fig 21 Morphological characteristics of capitula of male parent

图 22 梦幻系列绯红 F1 代单株
Fig 22 Morphological characteristics of single capitula of F1 hybrids

图 23 梦幻系列绯红 F1 代盛花期群体
Fig 23 Morphological characteristics of 'Meiguihong' in full-bloom stage

图 24 盆栽装饰
Fig 24 Zinnia elegans used for potted flower

图 25 盆花栽培
Fig 25 The dwarf varieties planted in the pots

图 01 '看月' 花朵形态
Fig 01 Flower morphology of *D. chinensis* 'Kanyue'

石竹
Dianthus chinensis

石竹（*Dianthus chinensis*），石竹科石竹属多年生草本植物，常作一二年生栽培。石竹是中国传统名花之一，原产中国西北、东北、华北、长江流域及东南亚地区，分布很广，除华南较热地区外，几乎中国各地均有分布。

品种介绍

1.'看月'（*D. chinensis* 'Kanyue'）

'看月'平均株高 25.7 cm，株幅 29.3 cm，簇生；茎直立，有节，多分枝；叶对生，叶长 3–7 cm，条形或线状披针形；花单朵或数朵簇生于茎顶，花径 5.5–6.1 cm，盛花期花量可达 60–75 朵 / 株，花瓣先端锯齿状，色彩由内到外依次成深紫、玫红和白色辐射状变化，呈现出极强的韵律和节奏美，微具香气。秋播次年春季 4 月份开花，播种期到盛花期约需 180 天左右；也可 2 月份春播，夏秋开花，播种期到盛花期约需 65 天左右，若栽培管理良好，4 月 –10 月均可开花，花期主要集中于 4–6 月；蒴果矩圆形或长圆形，种子扁圆形，黑褐色（图 01–05）。

生态习性

石竹性耐寒、耐干旱，不耐酷暑，夏季遇连阴多雨天气，多生长不良或枯萎，栽培时应注意遮荫降温。喜阳光充足、高燥，通风及凉爽湿润气候。要求肥沃、疏松、排水良好及含石灰质的壤土或沙质壤土，忌水涝，好肥。

主要园林用途

石竹常用于园林绿化之中，园林中可用于花坛、花境、花台或盆栽，也可用于岩石园和草坪边缘点缀。大面积成片栽植时可作景观地被材料，另外石竹有吸收二氧化硫和氯气的本领，凡有此气体污染的地方可以多种（图 06）。

繁殖技术要点

石竹是宿根性不强的多年生草本花卉，常作一二年生植物栽培。主要采用播种繁殖，也可进行分株和扦插繁殖。

播种繁殖：可采用春播和秋播，春播宜于 2 月下旬 –3 月上旬在温室内播种，秋播为 9 月中旬 –10 月中旬。春播和秋播繁殖技术要点相同，种子可点播或撒播；播种介质应排水良好、疏松，pH 值为 6.5–7.0，EC 值为 0.75，土壤事先进行消毒处理；种子具有嫌光性，播后需覆盖 0.5 cm 左右薄土，以不见种子为宜，置于避光处；播后浇透水，注意保湿；发芽适温为 20–25 ℃，一般秋播 5 天左右即可发芽出土。石竹播种后，当小苗长至 3–4 cm 时，将苗分栽至露天养护。经缓苗成活后，可少浇或不浇水，促进根系生长，然后加强水肥管理，石竹苗期生长的适温为 10–25 ℃。种苗进入快速生长期后，需要根据苗期长势，适当控制水分，并注意大环境通风，以防止病害。

扦插繁殖：多于 10 月至翌春 3 月进行，剪取老熟枝条，截成 6cm 左右长插穗，插于砂床或露地苗床，插后即灌溉遮光，长根后定植。

后期栽培养护同播种繁殖。

分株繁殖：多在花后利用老株分株，可在秋季或早春进行。后期栽培养护同播种繁殖。

栽培技术要点

石竹为阳性植物，生长、开花均需要充足的光照。栽培过程中苗长至 15 厘米高摘除顶芽，促其分枝，后注意适当摘除腋芽，不然分枝多，会使养分分散而生长变弱。

幼苗期：幼苗生长极快，容易徒长，待幼苗长至 2 对真叶时移栽，移至盆栽或露地栽培。移栽前要将穴盘从温室搬出，置于室外进行炼苗；移植宜在傍晚或阴天进行，以提高成活率。因其根系中侧根较少，移栽后恢复的速度较慢，应注意少伤侧根且不要窝根。定植前要进行土壤消毒并施足基肥，也可在定植后用敌克松 800 倍液灌根消毒。秋播（9 月）育苗时应注意避免正午太阳直射。在小苗上盆后，应给予全光照的环境条件，光照不足，容易引起营养生长旺盛，植株徒长，甚至影响开花时间。

营养生长期：生长期间宜放置在向阳、通风良好处养护，保持盆土湿润，约每隔 10 天左右施一次腐熟的稀薄液肥。夏季雨水过多，注意排水、松土。冬季宜少浇水，如温度保持在 5–8 ℃条件下，则冬、春不断开花。日常养护须注意水肥的控制，浇水要适度，过湿容易造成茎部腐烂，过干容易造成植株萎蔫。一般以植物叶片的轻度萎蔫为基准。施肥应掌握勤施薄肥，生长期可 0.2% 尿素和复合肥间隔施用，初花期只施复合肥，并适当拉长施肥时间。

石竹的病害主要是：立枯病、锈病和种蝇、蚜虫、蝼蛄等，可用药剂喷杀和毒饵诱杀。

图 02 '看月' 单株形态
Fig 02 Plant morphology of *D. chinensis* 'Kanyue'

图 03 看月 ' 花朵直径（5.5 cm）
Fig 03 The flower diameter(5.5 cm)of *D.chinensis* 'Kanyue'.

图 04 '看月' 植株株高（23 cm）
Fig 04 The plant height (23 cm) of *D. chinensis* 'Kanyue'

图 05 '看月' 群体照片
Fig 05 The population of *D. chinensis* 'Kanyue'

Chinese pink (*Dianthus chinensis*), a perennial herbaceous plant belongs to Caryophyllaceae, is usually used as annuals or biennial plants. Chinese pink is one of the Chinese traditional famous flowers, which is native to the Northwest, Northeast, North, Yangtze River basin and Southeast Asia of China. It has a wide distribution that nearly all over the China, except the south China for hotter climates.

Introduction of cultivars

1.'Kanyue'*(D. chinensis* 'Kanyue')

'Kanyue', the average plant height is 25.7 cm, plant width is 29.3 cm, and plant type is fascicles. The multiramose stem is upright with some nodes. The leaves are oppositifolious, the leaf length is 3–7 cm. The flower is single or several drifting on top of the stem, the flower diameter is 5.5–6.1 cm; the flower number of full–bloom stage can be 60–75 per plant. The petal of the flower is jagged , and the colors from inner to outer in turn are dark purple, rose and white, which present a strong rhythm beauty with slightly fragrant. It will blossom in April of next spring after sowing in autumn, for about 180 days from sowing to the full–bloom stage. What's more, it can also seed in February of spring; the time from sowing to full–bloom stage is about 65 days. It can blossom from April to October in a favorable cultivation and management, and the main flowering phase is from April to June. The shape of the capsule is oblong, and the seeds inside are oblate with black brown color (Fig.01–05).

Ecological habit

Chinese pink has excellent cold and drought tolerance, but intolerant of heat. If undergrowth continued cloudy and rainy weather in summer, the growth will be slow and even to die. In growth season, the cool shade measures should be paying more attention in cultivation. Chinese pink is like to grow in sunny, dry, ventilate and cool moist climate. And like fertile, loosened, well drained and calcareous loam or sandy loam. It should be avoided of waterlogging, and it prefers fertilizer.

Application

Chinese pink is often used for landscaping as a flower bed, flower border, flower table or pot culture, and it can also be used for ornament in rock gardens and lawn edge trimmer. When being planted extensively, it can be a ground cover plant for the landscape; moreover, Chinese pink has the ability to absorb sulfur dioxide and chlorine so can plant in the place where being contaminated by these gas (Fig.06).

Propagation

The mainly modes of reproduction of Chinese pink are seed propagation, and it can also use division and cutting propagation.

Seed propagation: It can adopt spring and autumn sowing. It is suitable to seed in late February or March of spring in the greenhouse, and autumn sowing date is from middle September to middle October. The key points of sowing: seeds can be bunched planted or broadcast sowed. The sowing matrix should be well-drained, loosen, the pH value is 6.5–7.0, the EC value is 0.75. The soil should be disinfected forehand. After seeding, cover 0.5 cm thin soil so as to unable to see the seeds, keep the seeds out of sunlight, then water enough, and keeping wet. The germination optimum temperature is 20–25 ℃. Generally speaking, it will sprout in 5 days. When the seedlings develop up to 3–4 cm after seeding, plant them dividually in the pot. After resuming growth, strengthen the management of water and fertilizer. The comfortable temperature of growth stage is 10–25 ℃. When the plants in a growth spurt, the water is controlled and pay attention to ventilation to prevent disease.

Cutting propagation: The cutting propagation usually proceeds from October to the next March. Cut the ripe old branches into 6 cm long cutting slips, then insert the cutting slips into the sand bed or outdoor seedbed, irrigate and shade the cutting slips after cutoff, and then plant in the pot.

Vegetative propagation: The vegetative propagation is often used for old plants which have blossomed; the division can be taken in autumn or early spring.

The key points of cultivation

Chinese pink is sun-like plant, it needs sufficient illumination in grow and blossom period. During the cultivation, excise the terminal bud to promote branching when the plants 15 cm high. After then paying attention to excise the axillary bud, otherwise, too many branches will disperse nutrient that the plants grow in firmly.

Seedling stage: When the seedlings have two or three pairs of leaves, transplant the seedlings into the pots. Move the whole aperture disk out of the greenhouse before transplanting, exercise seedlings in the outdoors for 1 or 2 weeks. Transplantation should be carried out in the evening or in cloudy days in order to increase the survival rate. In transplantation, pay attention to not injury laterals or not nest root as there is a little lateral root that the speed of recovery is relatively slow. Disinfect and fertilize the soil before transplanting. Or, using the 800 X Fenaminosulf irrigates the root after transplanting. Pay attention to avoid the midday direct solar radiation when autumn sowing in September. After potting, provide a full sun environment or else the vegetative growth will be too exuberant that the plants will be overgrow, what's more it will influence the flowering time.

Vegetative growth stage: The growing plants should be cultured in the place facing sun and drafty, keep the pots wet, fertilize the soil with low concentrated fertilizer about every ten days. Pay attention to drain and loosen the soil when it has excessive rainfall in summer. It needs less water in winter. If the temperature keeps in 5–8 ℃, the Chinese pink will blossom constantly. In the daily maintaining, pay attention to controlling the water and fertilizer. The stem will be rotten if it is too wet, and it will be wilted if it is too dry. In the growing stage, 0.2% urea and compound fertilizer are used to fertilize. In initial flowering, only compound fertilizer is used.

The main disease of Chinese pink: seedling blight, rust disease, aphid, and mole cricket and so on, spray the medicament or use the poison bait to prevention and cure.■

图 06 石竹在园林中的应用（拍摄于上海植物园）
Fig 06 The application of D.chinensis, taken in Shanghai botanical Garden

作者简介：

何燕红 / 女 / 讲师 / 博士 / 华中农业大学 / 湖北武汉
傅小鹏 / 女 / 副教授 / 博士 / 华中农业大学 / 湖北武汉
胡惠蓉 / 女 / 副教授 / 博士 / 华中农业大学 / 湖北武汉
叶要妹 / 女 / 教授 / 博士 / 华中农业大学 / 湖北武汉
刘国锋 / 男 / 教授 / 博士 / 华中农业大学 / 湖北武汉

Biography:

Yanhong He/female/Lecturer/Ph D/Huazhong Agricultural University/Wuhan, Hubei Province
Xiaopeng Fu/female/associate professor/Ph D/Huazhong Agricultural University/Wuhan, Hubei Province
Huirong Hu/female/associate professor/Ph D/Huazhong Agricultural University/Wuhan, Hubei Province
Yaomei Ye/female/professor/Ph D/Huazhong Agricultural University/Wuhan, Hubei Province
Guofeng Liu/male/professor/Ph D/Huazhong Agricultural University/Wuhan, Hubei Province

《世界园林》征稿启事
Notes to Worldscape Contributors

1 本刊是面向国际发行的主题性双语（中英文）期刊。设有作品实录、专题文章、人物／公司专栏、热点评论、构造、工法与材料（含植物）5个主要专栏。与主题相关的国内外优秀作品和文章均可投稿。稿件中所有文字均为中英文对照。所有投稿稿件文字均为Word文件。作品类投稿文字中英文均以1000-2500字为宜，专题文章投稿文字的中英文均以2500-4000字为宜。

2 来稿书写结构顺序为：文题（20字以内，含英文标题）、作者姓名（中国作者含汉语拼音，外国作者含中文翻译）、文章主体、作者简介（包括姓名、性别、籍贯、最高学历、职称或职务、从事学科或研究方向，现供职单位、所在城市、邮编、电子信箱、联系电话）。作者两人以上的，请注明顺序。

3 文中涉及的人名、地名、学名、公式、符号应核实无误；外文字母的文种、正斜体、大小写、上下标等应清楚注明：计量单位、符号、号字用法、专业名词术语一律采用相应的国家标准。植物应配上准确的拉丁学名。扫描或计算机绘制的图要求清晰、色彩饱和，尺寸不小于15cm*20cm；线条图一般以A4幅面为宜，图片电子文件分辨率不应小于300dpi（可提供多幅备选）。数码相机、数码单反相机拍摄的照片，要求不少于1000万像素（分辨率3872*2592），优先使用jpg格式。附表采用"三线表"，必要时可适当添加辅助线，表格上方写明表序和中英文表名，表序应于内文相应处标明。

4 作品类稿件应包含项目信息：项目位置／项目面积／委托单位／设计单位／设计师（限景观设计）／完成时间。

5 介绍作品的图片（有关设计构思、设计过程及建造情况和实景等均可）及专题文章插图均为jpg格式。图片请勿直接插在文字文件中，文字稿里插入配图编号，文末列入图题（须含中英对照的图号及简要说明）。图片文件请单独提供，编号与文字文件中图号一致。图题格式为：图01 xxx/Fig 01xxx。图片数量15-20张为宜。可标明排版时对图片大小的建议。

6 文稿一经录用，即每篇赠送期刊2本，抽印本10本。作者为2人以上，每人每篇赠送期刊1本，抽印本5本。

7 投稿邮箱 :Worldscape_c@chla.com.cn 联系电话 :86-10-51684910

---

1 Worldscape is an international thematic bilingual journal printed in dual Chinese and English. It covers five main columns including Projects, Articles, Masters / Ateliers, Comments, and Construction & Materials (including plants). The editors encourage the authors to contribute projects or articles related to the theme of each issue in both Chinese and English. All submissions should be submitted in Microsoft Word (.doc) format. Chinese articles should be 1000-2500 characters long. English articles should be 2500-4000 words long.

2 All the submitted articles should be organized in the following sequence: title (no more than 20 characters and the English title should be contained); author's name (for Chinese authors, pin yin of the name should be accompanied; for foreign authors, the Chinese translation of the name should be accompanied if applicable); main body; introduction to the author (including name, gender, native place, official academic credentials, position/title, discipline/research orientation, current employer, city of residence, postal code, E-mail, telephone number). For articles written by two or more authors, please list the names in sequence.

3 All persons, places, scientific names, formulas and symbols should be verified. The English submissions should be word-processed and carefully checked. Measuring units, symbols, and terminology should be used in accordance with corresponding national standards. Plants should be accompanied with correct Latin names. Scanned or computer-generated pictures should be sharp and saturated, and the size should be not less than 15cmx20cm. Diagrams and charts should be A4-sized. The resolution of digital images should be not less than 300dpi (authors are encouraged to provide a selection of images for the editors to choose from). The resolution of pictures generated by digital camera and digital SLR camera should be not less than 3872x2592, and .jpg formatted pictures are preferred. Annexed tables should be three-lined, and if necessary, auxiliary lines may be used. All tables should be sequenced and correspond to the text. Chinese-English captions should be contained.

4 All the submitted materials should be accompanied with short project information: site, area, client, design studio (atelier or company name), landscape designers (landscape architects) and completion date (year).

5 All project images (to illustrate the concept, design process, construction and built form) should be .jpg format. The images should be sent separately and not integrated in the text. All images should be numbered, and the numbers should be represented in the main body of the text. At the end of the text, captions and introductions to the images should be attached (Chinese-English bilingual text). The caption should be formatted as Fig 01 xxx. No more than 20 images should be submitted. Suggestions to image typeset may be attached.

6 The author of each accepted article will be sent 2 copies of the journal and 10 copies of the offprint. In the case of articles with 2 or more authors, each author will be sent 1 copy of the journal and 5 offprints.

7 Articles should be submitted to: Worldscape_c@chla.com.cn Tel:86-10-51684910.

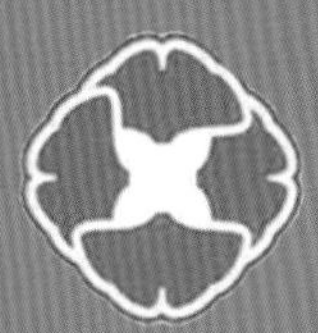

邳州市胜景银杏苗圃场

邳州市胜景银杏苗圃场位于苏北鲁南两大银杏树主产区中心，距京沪、京福高速入口十公里，310国道旁，地理环境优越，交通便利。我镇现存古银杏数万余株，果实累累，被评为国内银杏古树最多的地区，自二十世纪八十年代初，我镇由零星栽培，发展到成片大规模种植，基本是家户成形为农民带来可观的收入，给绿化工程带来美好的绿化景象。于92年大面积种植了银杏，多年来经过改良、移栽、培育出多种银杏嫁接树良种和绿化银杏苗木。